Beekeeping for Beginners

Your 360° Complete Guide From Hive to Honey: Master the art of beekeeping with a step by step journey

By

Anthony Greenwood

Copyright © 2024 by Anthony Greenwood.

All rights reserved. No part of this publication may be reproduced, distributed, or transmitted in any form or by any means, including photocopying, recording, or other electronic or mechanical methods, without the prior written permission of the publisher, except in the case of brief quotations embodied in critical reviews and certain other noncommercial uses permitted by copyright law.

Introduction **6**

   Why Keep Bees? 6

   Benefits to the Environment and You 7

   What to Expect from This Book 8

Chapter 1: Understanding Bees **10**

   1.1 The Life Cycle of Bees 10

   1.2 Understanding Bee Behavior and Social Structure 12

   1.3 Common Species of Honey Bees 14

Chapter 2: Getting Started with Beekeeping **16**

   2.1 Legal and Safety Considerations 16

   2.2 Choosing a Site for Beekeeping 18

   2.3 Equipment and Tools Needed 20

   2.4 Purchasing Your First Bees 24

Chapter 3: Setting Up Your Hive **26**

   3.1 Types of Hives: Pros and Cons 27

      3.1.1 Langstroth Hives 27

      3.1.2 Top-Bar Hives 28

      3.1.3 Warre Hives 29

   3.2 Step-by-Step Guide to Setting Up Your Hive 30

   3.3 Best Practices for Hive Placement 32

Chapter 4: Hive Management **34**

   4.1 Routine Inspections: What to Look For 34

   4.2 Feeding Your Bees: When and How 36

   4.3 Managing Pests and Diseases 38

   4.4 Seasonal Hive Management 40

Chapter 5: Understanding and Managing Swarms **42**

   5.1 What Causes Swarming? 42

   5.2 How to Prevent Swarming 44

5.3 What to Do If Your Bees Swarm 46

Chapter 6: Harvesting Honey 48

6.1 When to Harvest Honey 48

6.2 How to Harvest Honey Safely 50

6.3 Processing and Storing Your Honey 53

6.4 Other Hive Products: Wax, Propolis, and Royal Jelly 55

Chapter 7: Overwintering Your Bees 57

7.1 Preparing Your Hive for Winter 57

7.2 Winter Feeding and Care 59

7.3 Common Winter Challenges 61

Chapter 8: Advanced Topics in Beekeeping 63

8.1 Breeding Your Own Queens 63

8.2 Advanced Disease Management 65

8.3 Making Mead and Other Bee Products 67

Chapter 9: The Business of Beekeeping 69

9.1 Turning Beekeeping into a Business 69

9.2 Marketing and Selling Your Products 71

9.3 Networking with Other Beekeepers 73

Chapter 10: Sustainable Beekeeping Practices 75

10.1 Eco-Friendly Beekeeping Techniques 75

10.2 Minimizing Impact on Local Ecosystems 77

10.3 Promoting Biodiversity Through Beekeeping 79

Chapter 11: The Future of Beekeeping 82

11.1 Innovations in Beekeeping Technology 82

11.2 Challenges and Opportunities Ahead 84

11.3 Adapting to Climate Change 86

Chapter 12: Collect Your Bonus 89

We'd Love to Hear From You! *89*

Conclusions *90*

References *92*

Introduction

Why Keep Bees?

Hey there, future beekeeper! Are you buzzing with excitement about starting your beekeeping adventure? Great! Let's dive into why keeping bees might just be one of the best decisions you'll ever make.

Why beekeeping? Well, for starters, bees are fascinating creatures. Each bee in a hive plays a crucial role, from the hard-working worker bees to the queen bee who leads her colony. Beekeeping allows you to witness this natural wonder right in your backyard. Isn't it amazing to think about creating a space where you can observe and contribute to the magic of nature?

Now, let's talk about the environmental impact. Bees are critical pollinators. By keeping bees, you're helping to sustain the local flora. This isn't just good for your garden; it's vital for local agriculture and ecosystems. More bees mean more pollination, which helps in growing delicious fruits and veggies. So, how does it feel to be a hero for your local environment?

What about honey? Ah, yes, the sweet reward! Beekeeping can yield jars of fresh, natural honey. Imagine drizzling your own honey on pancakes, using it to sweeten tea, or soothing a sore throat. Plus, honey from your bees is all-natural, which makes it a healthier choice over processed sugars.

Curious about the community? Beekeepers are a friendly bunch! Getting into beekeeping can connect you with a community of passionate bee lovers. Whether it's local clubs or online forums, you'll find plenty of support and camaraderie. Ready to make some new friends who are as enthusiastic about bees as you are?

Thinking of the educational value? Whether you're a lifelong learner, a parent wanting to teach your kids about nature, or even a teacher, beekeeping is educational. You'll learn about biology, botany, chemistry, and more. Plus, it's hands-on, which makes learning fun and engaging.

How about the challenge? Beekeeping is both an art and a science. It challenges you to think critically, solve problems, and be patient. Each season brings new lessons as you manage your hive, care for your bees, and harvest your honey.

Interested yet? Before you zip up your bee suit and light your smoker, let's make sure you feel ready and excited to embark on this journey. Throughout this book, we'll explore each aspect of beekeeping in detail. So, stay tuned, ask questions, and get ready to be a part of your very own beekeeping adventure!

Suggestion for images: At this point in the book, a vibrant photo of a healthy, buzzing beehive in a lush garden would be ideal. This would not only illustrate the beauty of beekeeping but also help visualize the environment you might soon be creating.

What do you say? Are you ready to dive deeper into the world of beekeeping in our next section? Let's go!

Benefits to the Environment and You

Ready to explore how stepping into the world of beekeeping benefits not just your garden, but the planet and yourself? Let's get buzzing on this sweet journey!

For the Environment: Did you know that by becoming a beekeeper, you're stepping into the shoes of an environmental superhero? Bees play a critical role in pollinating plants. This isn't just about making your garden look pretty; it's about contributing to the health of local agriculture and natural ecosystems. Pollination by bees is crucial for about 75% of the plants we use for food and medicines. By keeping bees, you help ensure that plants reproduce and contribute to a stable, healthy environment. Imagine your bees visiting flower after flower, helping your entire community bloom beautifully!

For Biodiversity: Your bees will become local heroes, buzzing from plant to plant and increasing plant diversity and yields. This leads to a richer, more resilient ecosystem. By introducing a beehive to your area, you're creating a hub of activity that supports diverse plant and animal life. Plus, more diverse plant life means a more colorful and vibrant garden for you!

For Your Health: Now, let's talk about the sweet perks—honey! Harvesting your own honey means you get all the natural antioxidants, vitamins, and minerals that come with pure, raw honey. It's not just delicious; it's also a healthier alternative to refined sugar. Plus, spending time outside with your bees is a great way to relax and connect with nature, which is a natural stress reliever. Who knew that wearing a bee suit could be so calming?

For Learning and Fun: Beekeeping is a hands-on learning experience. You'll discover the intricate lives of bees and learn skills like carpentry (building hives), biology (understanding bee behavior), and chemistry (processing honey). It's a hobby that keeps you learning and growing.

For Community and Connection: Dive into the world of beekeeping, and you'll find yourself part of a warm and welcoming community. Whether it's local beekeeping clubs or online forums, you'll meet people who are eager to share their knowledge and experiences. It's a wonderful way to connect with others who share your interests.

So, are you ready to don your bee suit and embrace the benefits of beekeeping? Keep flipping the pages, and let's delve deeper into how you can get started on this amazing adventure. Your new bee friends are waiting! What do you think? Are you excited to see how your new hobby can make a big impact? Let's keep going and uncover more about the essential equipment and first steps in the next chapter!

What to Expect from This Book

As we gear up to explore the enchanting world of bees together, let's chat about what you can expect from this book. We've designed it to be your friendly guide, buzzing with information, tips, and hands-on activities that will transform you from a curious newbie into a confident beekeeper.

A Step-by-Step Journey: Starting with the basics, we'll walk you through understanding bees and their environment. Each chapter builds on the last, ensuring you grasp every aspect of beekeeping, from setting up your hive to managing and maintaining it throughout the year. You'll learn how to handle bees safely, monitor their health, and help them thrive in their new home.

Practical Advice and Tips: We know that the best way to learn is by doing, so expect plenty of practical advice. You'll find step-by-step instructions for daily and seasonal beekeeping tasks. Plus, we'll share troubleshooting tips for common problems, ensuring you're prepared for anything your beekeeping journey throws at you.

Engaging Q&A Sections: To make your reading experience more interactive, we've included Q&A sections throughout the book. These will help you review key concepts and apply what you've learned. It's like having a beekeeping mentor right at your fingertips!

Beautiful Illustrations and Photos: Since seeing is believing (and understanding!), we've packed this book with beautiful illustrations and photos. These visuals will help you identify different types of bees, understand the structure of hives, and follow the step-by-step procedures correctly. We suggest using vibrant images of beekeeping in action to make the content come alive.

Environmental Impact Awareness: Beyond just keeping bees, this book emphasizes the broader environmental impacts of beekeeping. You'll learn how your activities as a beekeeper can support local ecosystems and contribute to global conservation efforts.

By the end of this book, you'll not only have a deep appreciation for these incredible pollinators but also the practical skills to manage your own hives successfully. Are you ready to get started? Flip the page, and let's begin this buzz-worthy journey together!

Chapter 1: Understanding Bees

As we embark on this buzzing journey, you'll dive into the fascinating world of bees, learning about their life cycles, behavior, and the diverse species you might encounter. This chapter is designed to lay the foundation for your beekeeping adventures, providing you with a solid understanding of these incredible insects.

1.1 The Life Cycle of Bees: Discover the amazing journey from egg to adult, and learn how each stage of a bee's life plays a critical role in the colony's health and productivity.

1.2 Understanding Bee Behavior and Social Structure: Delve into the complex social life of bees, exploring how they communicate, work together, and organize themselves within the hive.

1.3 Common Species of Honey Bees: Get acquainted with the different species of honey bees, each with unique traits and qualities that can influence your beekeeping experience.

By the end of this chapter, you'll not only have a deeper appreciation for these little pollinators, but you'll also be equipped with the knowledge to start your beekeeping journey on the right foot. Ready to learn about these extraordinary creatures? Let's buzz right in!

1.1 The Life Cycle of Bees

Ready to unravel the mysteries of a bee's life? Well, strap on your wings—we're about to embark on a fascinating journey from tiny egg to busy bee. Understanding the life cycle of bees is not just interesting, it's crucial for anyone looking to step into the shoes of a beekeeper.

Egg-cellent Beginnings: Our journey starts in the heart of the hive, where the queen bee lays her eggs. Picture this: a tiny, white, sausage-shaped egg, no larger than a grain of rice. Each egg is delicately placed in its own cell within the honeycomb. Here, in the cozy confines of the hive, the egg stage lasts about three days. Isn't it egg-citing to think about the potential packed into such a small package?

The Larval Stage: As the eggs hatch, they reveal larvae, tiny white grubs that are voracious eaters. The hive's worker bees spring into action, feeding these hungry babies a diet of royal jelly, pollen, and honey. This menu helps the larvae grow rapidly, shedding their skin multiple times. Over about five to six days, these larvae bulk up, preparing for the next big change. During this stage, the hive is abuzz with the sounds of communal care—truly a testament to the teamwork in bee society!

Pupa, the Transformation Chamber: Once fully grown, the larvae spin themselves into silky cocoons, transitioning into the pupal stage. This is where the magic happens! Inside their protective coverings, the larvae metamorphose into the bees we recognize: wings, antennae, legs, and all. This process takes about 12-14 days, a period of intense transformation hidden from view. If we could peek inside, we'd see a marvelous natural spectacle of development and change.

Birth of a Bee: After about 21 days from egg to emergence, the adult bees chew their way out of the wax caps sealing their cells. They are born ready to work, each with a role to play, depending on their type—worker, drone, or future queen. The first job of these newly emerged workers is to clean their own birth cells, proving that in the bee world, cleanliness is next to bee-liness!

Roles and Responsibilities: The adult stage is where the diversity of bee life truly shines. Worker bees start off performing chores inside the hive, such as feeding the larvae and tending to the queen. As they age, their roles shift to guarding the hive and finally, venturing out as foragers. Drones, on the other hand, have just one goal: mating with a queen. And the queen? Her majesty's life revolves around laying the eggs that will sustain the colony's future.

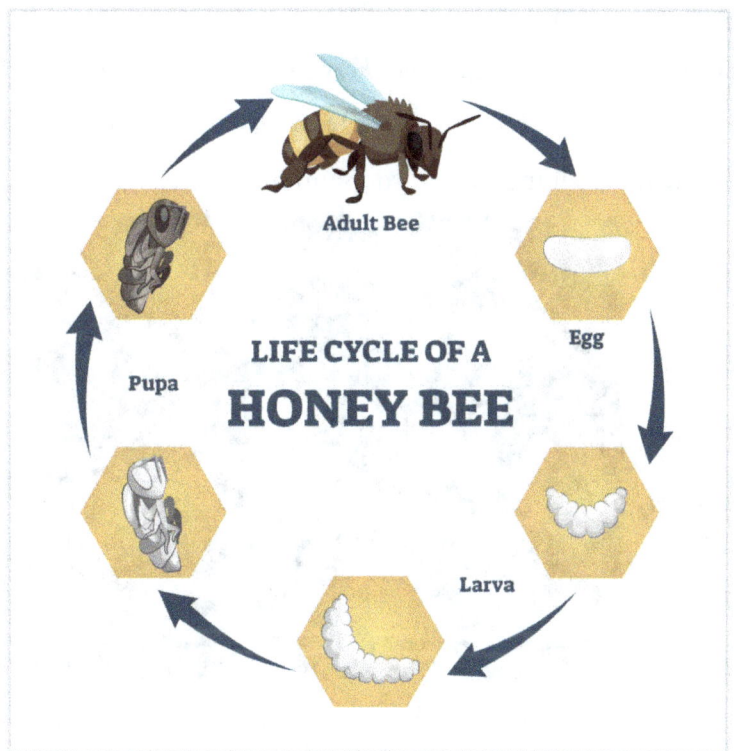

By understanding the life cycle of bees, you're not just learning how bees grow; you're gaining insight into the delicate balance of the hive and the essential roles each stage plays in the life of this incredible community. So, as we buzz forward, keep these stages in mind—they're the building blocks of your future buzzing empire! Are you ready to explore more about these remarkable creatures and their behaviors in the hive? Let's keep the curiosity flowing as we move on to the fascinating social structure of bees in the next section!

1.2 Understanding Bee Behavior and Social Structure

This part of our journey explores the incredibly organized and complex society that bees create and maintain. It's a fascinating system that rivals the most sophisticated civilizations humans have built!

The Hive Mind at Work: When we talk about bee behavior, it's essential to understand the concept of the 'hive mind.' This doesn't mean that each bee isn't an individual, but rather that they work so harmoniously that the hive functions like a single organism, with each bee playing a specific role for the greater good of the community. How cool is that?

Roles in the Hive: In the world of bees, everyone has a job to do, and each role is critical to the hive's success:

The Queen: The queen is the heart of the hive. Her main role? Laying the eggs that will spawn the hive's next generation of bees. But she's more than just a baby-maker; her chemical signals, or pheromones, help regulate the life of the hive, maintaining social order and productivity.

Worker Bees: These are all female and are the most populous in the hive. They are the true heroes, taking on the most tasks, from nursing the young to foraging for food, building and repairing the hive, and protecting it. As workers age, their jobs shift from inside duties to the more hazardous outside work—a real bee's life cycle of career development!

Drones: The male bees, known as drones, have one primary purpose: to mate with a new queen. They may seem like the privileged few, but life is tough for a drone. After fulfilling their role in mating, they are typically expelled from the hive to preserve precious resources.

Communication Is Key: Bees communicate in remarkable ways. One of their most famous methods is the 'waggle dance.' Through this dance, bees can share information about the direction and distance to sources of food with their hive mates. Imagine finding out where the nearest pizza place is just by watching someone dance!

Social Order and Harmony: Bee society is all about harmony and efficiency. Order is maintained not through conflict but through pheromones and the bees' innate understanding of the common good. This ensures that the hive runs smoothly and every bee contributes effectively without oversight. It's a model of cooperation and community that humans could learn a lot from!

Defense and Protection: When it comes to protecting their home, bees are not to be underestimated. Their coordinated defense strategies can fend off predators much larger than themselves. It's a remarkable demonstration of what unity and determination can achieve.

Understanding these aspects of bee behavior and social structure not only makes beekeeping more engaging but also equips you with the knowledge to better manage and care for your bees. You're not just keeping bees; you're participating in a complex ecological system where each member has a vital role.

1.3 Common Species of Honey Bees

Not all honey bees are created equal, and knowing the differences can help you choose the best fit for your beekeeping goals and local environment. Let's take a flight through the fascinating world of the most common honey bee species.

1. The Western Honey Bee (Apis mellifera): This is the superstar of the beekeeping world, widely known and loved for its ability to produce copious amounts of honey and its adaptability to various climates and environments. The Western honey bee is a social creature with a complex behavioral repertoire and a highly structured social order. They are the most common bees kept by beekeepers globally and are prized for their efficiency and relatively gentle nature.

2. The Eastern Honey Bee (Apis cerana): Often found in the wilds of Asia, the Eastern honey bee is smaller than its Western cousin and is known for its slightly more docile nature. This species is particularly interesting because of its ability to fight off certain pests and diseases better than the Western honey bee, including the dreaded Varroa mite—a major pest for honey bees worldwide.

3. The Giant Honey Bee (Apis dorsata): True to its name, the Giant honey bee builds large, impressive single-comb hives in the open air. These bees are found primarily in forested areas of Southeast Asia. While not typically managed by beekeepers due to their aggressive nature and tendency to nest high up in trees, they are fascinating for their migratory behaviors and the large quantities of honey they can produce.

4. The Dwarf Honey Bee (Apis florea): Another species native to Asia, the Dwarf honey bee is smaller than most of its relatives and builds open-air nests similar to those of the Giant honey bee, but much smaller. These bees are known for their gentle nature and are less commonly kept due to their smaller size and less honey production.

Each of these species has unique traits that can affect their care, handling, and the benefits they bring to your beekeeping adventure. Understanding these differences is key to managing your hives effectively.

Why does it matter? Choosing the right species of honey bee can impact everything from the type of honey produced to how well the bees thrive in your local environment. For example, if you live in a region where Varroa mites are a significant problem, you might opt for Apis cerana over the more common Apis mellifera.

Bee-ware of Challenges: Each species also comes with its own set of challenges. The more you know about these bees, the better prepared you'll be to handle any issues that come your way. For instance, while the Giant honey bee can produce a lot of honey, their aggressive nature and high nesting sites make them less ideal for a backyard beekeeper.

Chapter 2: Getting Started with Beekeeping

Starting from this Chapter, your beekeeping journey truly begins to take shape! Here, we're going to equip you with all the essentials you need to start your beekeeping adventure with confidence. From understanding legal requirements to choosing the perfect site, and gathering the right tools to buying your very first bees, this chapter covers all the bases to ensure you start off on the right foot.

2.1 Legal and Safety Considerations: Before you dive into beekeeping, it's crucial to understand the legal landscape and safety protocols. This section will guide you through navigating local laws and regulations, helping you keep your beekeeping dream compliant and safe for everyone involved.

2.2 Choosing a Site for Beekeeping: Selecting the right location is key to your success. We'll explore what makes a site ideal for beekeeping, considering factors like exposure to sunlight, wind protection, and proximity to water and foraging sources.

2.3 Equipment and Tools Needed: Get ready to gear up! Here, we'll break down the essential equipment and tools every beekeeper needs. From hives to protective clothing, this section ensures you have everything required to care for your bees properly.

2.4 Purchasing Your First Bees: Finally, the exciting part—bringing home your bees! We'll discuss how to choose healthy bees, where to buy them, and tips for safely introducing them to their new home.

By the end of this chapter, you'll be well-prepared to establish your first colony and start contributing to the fascinating world of beekeeping. Let's get buzzing on setting up your hive and embarking on this incredible adventure!

2.1 Legal and Safety Considerations

Before you set up your hives and don your bee suit, it's crucial to buzz through some important legal and safety considerations. Beekeeping is a wonderful hobby and can even turn into a buzzing business, but like all great adventures, it starts with understanding the rules and how to keep everyone safe—including you and your bees!

Know the Law: Beekeeping laws vary widely depending on where you live. Some areas welcome beekeepers with open arms, while others have strict regulations or require specific permits. Why? Because bees are vital to our ecosystem, but they need to be managed responsibly to ensure they don't become a nuisance or hazard to people, especially in urban areas.

- **Local Ordinances:** Start by checking with your local government. There might be zoning laws that specify where hives can be located, how many hives you can have, and how far they need to be from property lines or public spaces.
- **Registration:** Many places require beekeepers to register their hives. This helps local authorities keep track of bee populations and ensure beekeepers are managing their hives responsibly.
- **Health Regulations:** Some jurisdictions have health regulations to prevent diseases that can spread through bee colonies and potentially affect local agriculture.

Keeping it Safe: Safety is paramount in beekeeping—not just for you but for your neighbors and the bees themselves.

- **Protective Gear:** Always wear the right protective gear. A bee suit, gloves, and a veil aren't just fashion statements—they're essential tools to protect you from stings.
- **Bee Management:** Learn how to manage your bees effectively. This means understanding bee behavior, knowing how to handle bees gently and calmly, and recognizing signs of distress or disease in your hive.
- **Allergy Checks:** Be aware of allergies—both your own and others'. Bee stings can cause severe allergic reactions in some people. If you or a family member are allergic to bee stings, you'll need to take extra precautions.

Public Safety: Part of being a responsible beekeeper is ensuring that your bees don't become a problem for others.

- **Swarm Control:** Swarms can be alarming to the uninitiated. Learn how to prevent and control swarming to minimize risks to the public.
- **Communication:** Keep your neighbors informed. Many people are understandably nervous about living near bees. Educate them about what you're doing to keep the area safe, and what they can expect from having bees as neighbors.

Sustainable Practices: Adopting sustainable and ethical beekeeping practices is not only good for the bees but also for the environment and your community.

- **Pest Management:** Use environmentally friendly methods to manage pests and diseases within your hives. Harsh chemicals can harm bees and the products they produce, like honey and wax.
- **Water Sources:** Provide water for your bees. This keeps them from venturing into neighbor's yards in search of moisture, which can be a common source of conflict.

Insurance: Consider getting insurance. Yes, beekeeping can be risky, and having liability insurance might protect you if someone is stung or property is damaged because of your bees.

By understanding and adhering to these legal and safety guidelines, you'll not only protect yourself and your bees, but you'll also ensure that your beekeeping practice contributes positively to your community. Ready to move on and pick the perfect spot for your hives? Let's fly into the next section!

2.2 Choosing a Site for Beekeeping

Picking the right spot for your hives is like choosing a new home for your family—it needs to be just right. The location of your hives can make a big difference in your bees' health and productivity. So, let's get our maps out and find the perfect spot for your buzzing buddies!

Sunlight and Shade: Bees thrive in a spot that catches the morning sun. This early light helps them start their day's work of foraging and hive maintenance. However, they also need protection from the harsh midday sun. A location that offers morning sunlight with afternoon shade is ideal. Why not plant some tall flowers or shrubs that provide natural afternoon shade? Not only will they help regulate the temperature, but they'll add some beauty to your bee garden too!

Wind Protection: Bees aren't big fans of windy spots, which can chill them and make flying difficult. Choose a site protected from strong winds, perhaps by a natural windbreak like a hedge or a row of trees. This will keep your bees cozy and stop the wind from cooling their hive too much, especially during the colder months.

Water Source: Bees need water, not just for drinking but also to help regulate the temperature of their hive. A nearby natural water source is a treasure, but make sure it's not too close to avoid dampness around the hive, which can lead to diseases. A simple birdbath or a shallow water dish with stones for bees to land on can also do the trick!

Flight Path: Consider the flight path your bees will take as they zoom in and out of their hive. Keep hives away from high-traffic areas like paths and especially entrances to your home. You want to give your bees a clear flight path to minimize the chances of bee-human collisions.

Stability and Ground Conditions: The ground where you place your hive should be stable and well-drained. You don't want your hive toppling over in a strong gust of wind or sitting in a puddle after a rainstorm. If necessary, consider raising your hives off the ground with a sturdy stand.

Accessibility: You will need regular access to your hives for maintenance and harvesting, so don't put them somewhere too hard to reach. Make sure you can safely carry equipment and harvest your honey without having to trek through tough terrain.

Safety and Privacy: Last but not least, think about the safety and privacy of both your bees and your neighbors. Keep hives at a respectful distance from property lines to avoid any conflicts with neighbors who might not be as enthusiastic about bees as you are. Also, consider how visible the hives are to avoid attracting too much curiosity, which could disturb the bees.

Choosing the right site for your bee hives is not just about finding a spot; it's about creating a harmonious environment where your bees can prosper and play their part in the natural world. With these tips, you're well on your way to becoming a thoughtful and successful beekeeper. Ready to gear up? Let's buzz into the tools and equipment you'll need to start your beekeeping adventure!

2.3 Equipment and Tools Needed

Just as a knight needs armor and a painter needs brushes, a beekeeper needs the right tools to ensure the job is done safely and effectively. Let's buzz through the essential equipment and tools you'll need to start your beekeeping journey. Get ready to equip your beekeeping kit with all the must-haves to keep your bees buzzing happily.

1. Beehive – Your Bees' New Home:

The heart of your beekeeping endeavor is the beehive, but not just any box will do! You'll need a well-constructed hive that can house your bees comfortably and protect them from the elements. There are several types of hives to choose from:

- **Langstroth Hive:** The most popular choice, consisting of vertically stacked rectangular boxes with frames for the bees to build their honeycombs.
- **Top-Bar Hive:** A horizontal hive where bees build their comb hanging from removable bars, great for natural beekeeping practices.
- **Warre Hive:** Designed to mimic the bees' natural living conditions, encouraging natural hive behaviors.

Each type has its pros and cons, depending on your beekeeping style and goals. Consider what will work best for your environment and handling preferences.

2. Protective Gear – Suit Up!

Safety first! Beekeeping can be a sticky (and stingy!) business without the right protective wear:

- **Bee Suit:** A full-body suit that covers you from head to toe, made of light-colored material to deter bees (who are attracted to dark colors).

- **Gloves:** Thick gloves that protect your hands but still allow you to handle your tools and frames delicately.

- **Veil and Hat:** A veil is crucial to protect your face and neck from stings. Attached to a hat, it keeps the veil away from your face, providing visibility and protection.

3. Smoker – Keeping the Peace:

A smoker is essential for calming your bees when you need to work on the hive. Smoke mimics the signals of a forest fire, prompting bees to consume honey in preparation for a potential move, which makes them less aggressive and easier to handle.

4. Hive Tools – The Beekeeper's Toolkit:

These are the must-have tools for any beekeeper:

- **Hive Tool:** A versatile metal device used for prying apart hive bodies and scraping wax and propolis off of the hive parts.

- **Frame Gripper:** Helps in lifting the frames out. Beekeeping can get slippery with all that honey and wax!

- **Uncapping Fork:** Useful for honey harvesting, this tool helps remove the wax caps from honeycomb cells.

5. Feeders – Feeding Your Fuzzy Friends:

Especially important when natural food sources are scarce, feeders help you provide sugar syrup or other nutrients to sustain your colony.

- **Entrance Feeder:** Fits into the hive entrance and is easy to monitor and refill.
- **Top Feeder:** Sits on top of the uppermost hive box, allowing bees easy access without much disturbance.

6. Brush – Gentle Bee Handling:

A soft-bristled brush is gentle enough to coax bees off the comb, or clear bees from areas where you need to work. It's an indispensable tool during honey harvests or when performing hive inspections.

7. Books and Resources – Knowledge is Power:

Lastly, don't forget to arm yourself with knowledge! A good beekeeping book can be an invaluable resource, offering insights and tips that can save you a lot of trouble down the line.

By ensuring you have these essentials, you're setting yourself up for success in the beekeeping world. Each tool and piece of equipment plays a pivotal role in managing your hive effectively, keeping your bees healthy, and ensuring your safety as you delve into the rewarding practice of beekeeping. So, gear up, it's time to get buzzing! Ready to purchase your first bees? Let's move on to how and where to find your new buzzing companions!

2.4 Purchasing Your First Bees

The moment has buzzed its way here—you're ready to bring home your first colony of bees! This step is as thrilling as it is vital. Picking the right bees is crucial to your success as a beekeeper, so let's navigate through the process of selecting and bringing home your buzzing companions.

Choosing the Right Type of Bees: Before you even consider where to buy your bees, it's important to decide what type of bees you want. Different species and strains have varying characteristics and temperaments, which can impact their suitability for your climate and beekeeping style.

- **Italian Bees (Apis mellifera ligustica):** Known for their gentle nature and prolific honey production, they are ideal for beginners.
- **Carniolan Bees (Apis mellifera carnica):** Appreciated for their ability to handle colder climates and gentle temperament.
- **Russian Bees:** Valued for their resistance to certain pests and diseases, these might require a bit more beekeeping savvy due to their tendency to swarm.

Where to Buy Your Bees: You have a few options when it comes to acquiring your bees, and each comes with its pros and cons.

- **Local Beekeeping Clubs or Associations:** Purchasing bees from a local source can have huge benefits, including acclimated bees that are likely to thrive in your environment.
- **Reputable Apiaries or Bee Farms:** Ensure the apiary is reputable and can provide health certifications for their bees, guaranteeing that you're starting with a healthy colony.
- **Online Retailers:** While convenient, buying bees online requires thorough research to ensure the bees come from a reputable supplier.

Nucleus Colony vs. Packaged Bees:

- **Nucleus Colony:** Often referred to as a "nuc," this is a small, established colony with its own queen, workers, brood, and food stores. It's a bit more expensive but generally easier and quicker to establish.
- **Packaged Bees:** These are typically a box containing several thousand bees and a separate caged queen. This option can be more challenging and takes longer to establish but is often less expensive.

Bringing Your Bees Home: Transporting bees requires care to ensure their safety and reduce stress. If you're transporting them in a vehicle, ensure good ventilation and a cool, dark environment to keep the bees calm.

Setting Up Your New Hive: Upon arrival, your new bees will need to be introduced to their new home—a process that varies slightly between nucs and packaged bees.

- For nucs, it's often as simple as transferring the frames into your hive.

- For packaged bees, you'll need to carefully release the bees into the hive and ensure the queen is safely introduced.

Aftercare: Once your bees are in their new home, monitoring them is crucial. Ensure they have enough food (supplement with sugar syrup if needed), check that the queen is out of her cage and laying, and watch for any signs of distress or disease.

By taking these steps, you ensure that your first foray into beekeeping gets off to a buzzing start. Remember, the more care you take in selecting and introducing your bees to their new environment, the more successful your beekeeping venture will be. Now that your bees are settled, let's prepare for the exciting world of hive management and bee care! Ready to see your new colony thrive? Let's keep the buzz going!

Chapter 3: Setting Up Your Hive

This crucial chapter will guide you through everything you need to know about selecting and setting up your hive. We'll dive into the various types of hives, each with its own set of advantages and challenges, ensuring you find the perfect home for your bees.

3.1 Types of Hives: Pros and Cons

We'll explore the three popular types of hives—Langstroth, Top-Bar, and Warre—each uniquely suited to different beekeeping styles and goals. Whether you're looking for ease of maintenance, natural beekeeping methods, or something that fits in a small space, this section will help you make an informed decision.

- **3.1.1 Langstroth Hives:** Discover why this is the most widely used hive type and how its design aids in bee management and honey production.
- **3.1.2 Top-Bar Hives:** Learn about the simplicity and natural approach of the Top-Bar hive, which promotes bee health and makes honey extraction less stressful for the bees.
- **3.1.3 Warre Hives:** Understand the philosophy behind the Warre hive's design, which aims to mimic the bees' natural living conditions, potentially reducing stress and disease.

3.2 Step-by-Step Guide to Setting Up Your Hive

This section provides a detailed walkthrough of the setup process, from assembling your hive to ensuring it's ready for your bees. With step-by-step instructions, you'll feel confident setting up your hive correctly, ensuring a strong start for your colony.

3.3 Best Practices for Hive Placement

Finding the right location for your hive can significantly affect your bees' health and productivity. We'll cover the key factors to consider, such as sun exposure, wind protection, and safety from predators, to help you choose the ideal spot in your garden or apiary.

By the end of this chapter, you'll be fully equipped to choose and set up your hive, placing it in the perfect location to maximize the health and happiness of your bees. Let's get buzzing and build a thriving home for your new bee colony!

3.1 Types of Hives: Pros and Cons

Embarking on the beekeeping journey begins with selecting the right hive. Each type of hive offers unique benefits and challenges, tailored to different styles of beekeeping. Understanding the specifics of each can greatly enhance your beekeeping experience. Here's a detailed breakdown of the three popular types of hives: Langstroth, Top-Bar, and Warre.

3.1.1 Langstroth Hives

Developed by Lorenzo Lorraine Langstroth in the 1850s, the Langstroth hive is revered for its revolutionary "bee space" and removable frames, which allow beekeepers to inspect and manage their hives with minimal disruption to the bees.

- **Pros:**
  - **Ease of Inspection and Management:** The removable frames make it easy to inspect each frame for health and productivity, manage diseases, and harvest honey without destroying the hive structure.
  - **Scalability:** These hives can be easily expanded or contracted by adding or removing boxes, making it simple to adjust based on the colony's needs.
  - **High Honey Yields:** Due to their efficient structure, Langstroth hives typically produce more honey than other types.
- **Cons:**
  - **Weight:** The boxes, especially when full, can be heavy and cumbersome, potentially requiring multiple people to handle.
  - **Space Requirement:** These hives consume more space, which might not be ideal for smaller yards or gardens.

3.1.2 Top-Bar Hives

A favorite in natural beekeeping circles, the Top-Bar hive mimics the natural formation of combs. Bees build their honeycomb hanging from bars across the top of a horizontal hive.

- **Pros:**
    - **Ergonomic Management:** All activities are done at waist level, which reduces strain on the beekeeper's back and makes the hive easier to handle.
    - **Natural Bee Behavior:** Bees build comb naturally without foundation, which can lead to healthier bees and more natural bee behavior.
    - **Less Expensive:** Generally lower startup costs as they require fewer materials and accessories than Langstroth hives.
- **Cons:**
    - **Lower Honey Production:** They typically yield less honey due to the limited space for bees to store honey.
    - **Comb Fragility:** The natural comb is not reinforced by frames, making it more susceptible to damage during inspections.

3.1.3 Warre Hives

Named after Abbé Émile Warré, the Warre hive is designed to be a low-maintenance, vertical top-bar hive that aims to mimic the natural environment of a bee colony as closely as possible.

- **Pros:**
    - **Low Maintenance:** Designed to be minimally invasive, it's ideal for "leave-it-be" beekeepers who prefer to interfere as little as possible.
    - **Natural Environment:** Like Top-Bar hives, Warre hives allow bees to build their comb naturally, which can help reduce stress and disease.
    - **Insulation:** Typically better insulated than other hives, making them suitable for colder climates.
- **Cons:**
    - **Honey Harvesting Difficulty:** Accessing and harvesting honey can be more challenging and disruptive, as it usually involves removing whole boxes or cutting comb.
    - **Less Common:** Fewer beekeepers use Warre hives, which can make finding resources and community support more difficult.

By understanding the strengths and limitations of each hive type, you can choose the best fit for your beekeeping goals, environment, and personal capacity. This knowledge will help you set up a thriving home for your bees, tailor-made to encourage their health and productivity. Whether you aim for high honey production, ease of management, or a more natural approach, there's a hive type just right for your beekeeping aspirations.

3.2 Step-by-Step Guide to Setting Up Your Hive

Alright, buzzing pals! It's time to roll up your sleeves and get your hives set up. Setting up your hive correctly is crucial for the well-being of your bees and the success of your beekeeping adventure. Let's walk through the steps, ensuring that you create a buzzing haven for your new friends.

Step 1: Gather Your Materials Before you begin, make sure you have all the necessary materials and tools at hand. You'll need your hive components (whether it's a Langstroth, Top-Bar, or Warre), protective gear, a hive tool, screws or nails, a hammer or screwdriver, and possibly some wood glue or sealant to ensure everything fits snugly and stands up to the elements.

Step 2: Choose the Ideal Location Your bees' productivity and health heavily depend on where you place their hive. Pick a spot that gets morning sun and afternoon shade, is shielded from strong winds, and isn't in a damp area. The location should be flat and stable to prevent the hive from tilting or toppling over. Remember, your hive should be accessible for regular checks and maintenance, yet away from heavy foot traffic to keep both the bees and passersby safe.

Step 3: Assemble the Hive Stand Start by assembling the hive stand if you are using one. A good stand not only keeps the hive off the ground away from moisture and pests but also makes it easier to work with the hive by raising it to a more comfortable level. Ensure the stand is level and stable.

Step 4: Set Up the Hive Body Whether you're using a Langstroth, Top-Bar, or Warre hive, the principle remains the same—start from the bottom and work your way up.

- For Langstroth hives, place the bottom board first, then stack the brood boxes, and add the supers for honey storage above them. Ensure each box fits snugly on top of the other to prevent any gaps.
- For Top-Bar hives, ensure the bars are evenly spaced at the top of the hive body. The bars are where the bees will build their comb, so they need to be aligned correctly.
- For Warre hives, start with the bottom box and add additional boxes underneath as the bees build comb and fill the upper boxes.

Step 5: Install Frames or Bars Insert the frames or bars into your hive. In a Langstroth hive, the frames should hang parallel to each other within the boxes. For Top-Bar and Warre hives, place the bars at the top of the hive sections. If using foundation (wax sheets that guide comb building), fit these into your frames unless you are going for a foundationless approach.

Step 6: Add the Queen Excluder (if applicable) For Langstroth hives, you might choose to use a queen excluder between the brood boxes and the honey supers. This prevents the queen from laying eggs in the honey storage area but allows worker bees to move up and store honey.

Step 7: Place the Inner and Outer Covers Top off your hive with the inner cover and then the outer lid or telescoping cover, which will protect your bees from the elements. Ensure the lid is secure but has proper ventilation to prevent moisture buildup inside the hive.

Step 8: Set Up Perimeter Protection Consider setting up a barrier or fence around your hive, especially if you're in an area with bears or other large animals. Electric fences can be very effective in keeping predators out.

After setting up your hive, give it a day or two before introducing your bees to their new home. This allows any strong smells from new materials to dissipate and the hive to settle into its environment.

With your hive now set up, you're ready to welcome your bees into their new home. Let's ensure their transition is as smooth as your setup process was! Onward to introducing your bees to their new abode and beginning your journey as a beekeeper with a solid foundation—literally!

3.3 Best Practices for Hive Placement

Now that you've got your hive ready, it's time to find the perfect spot in your backyard or garden. Placing your hive thoughtfully is crucial for the success and health of your bee colony. Let's explore some best practices for hive placement that ensure your bees are happy, healthy, and productive.

Sunshine and Shade: Bees are like goldilocks; they need conditions that are just right. Your hive should catch the early morning sun, which encourages bees to start their day early. However, too much afternoon sun can overheat the hive, so a little afternoon shade can be beneficial. Positioning your hive so it faces southeast often captures this balance perfectly. This setup encourages bees to wake up early and keeps the hive warm without the harsh midday heat.

Wind Protection: Bees aren't fans of windy conditions, which can make flying difficult and chill their home. To protect them, place your hive near a natural windbreak like a fence, wall, or hedge. This barrier should shield the hive from prevailing winds, especially during colder months, but shouldn't block airflow completely. Good air circulation is essential to keep humidity levels within the hive in check.

Dry and Stable Ground: The ground under a beehive should be stable and well-drained. Wet conditions can lead to mold and disease in the hive. Ensure the area doesn't collect standing water, and consider elevating the hive slightly with a sturdy stand. This not only prevents moisture-related issues but also makes it harder for ground pests to enter the hive.

Keep it Quiet and Calm: Bees thrive in peaceful environments. Avoid placing hives near busy paths, roads, or loud areas. The constant vibration and noise can stress the bees, which may lead to aggressive behavior and reduced productivity. A quiet spot helps to keep the bees calm and less likely to become stressed.

Safety for All: Consider the safety of both the bees and the people around them. Keep the hive out of high-traffic areas where people and pets frequent to minimize disturbances and the risk of stings. Also, ensure that the hive entrance faces away from your home, walking paths, or public areas to direct bee traffic where it won't cause issues for neighbors or guests.

Accessibility for Maintenance: You'll need to check on your hives regularly, so place them where they are easily accessible for you. Think about how you'll reach the hive with all your equipment during inspections or harvests. You shouldn't have to trek through difficult terrain or dense vegetation to get to your bees.

Regulations and Respect: Always remember to check local regulations regarding beekeeping, and maintain good relationships with your neighbors. Keeping your hives at a considerate distance from property lines and discussing your beekeeping plans with neighbors can prevent misunderstandings and complaints.

Visual Checkpoints: Finally, when you think you've found the perfect spot, observe it at different times of the day. This observation will help you confirm that conditions remain ideal, considering sun, wind, and activity around the area throughout a typical day.

By following these best practices, you ensure that your hives are placed in the best possible location for the bees to prosper. A well-placed hive leads to happier bees and a more fruitful beekeeping experience. Ready to move on to welcoming your bees into their new home? Let's make sure their introduction is as smooth as honey!

Chapter 4: Hive Management

In this chapter we'll dive into the essential practices of hive management to ensure your bee colony thrives throughout the year. Managing a hive goes beyond simply housing bees; it involves regular inspections, proper feeding, disease control, and adapting to seasonal changes. Here's what you'll learn in this chapter:

4.1 Routine Inspections: What to Look For

Discover how to conduct thorough inspections, understanding what's normal and what might be a sign of trouble. Learn to identify healthy brood patterns, signs of queen presence, and ensure adequate food stores and space for your bees.

4.2 Feeding Your Bees: When and How

We'll guide you through the best practices for feeding your bees, including what types of feed to use, how to feed them safely, and determining when supplementary feeding is necessary to support your hive.

4.3 Managing Pests and Diseases

Protect your bees from common threats by learning about preventative measures and treatments for pests and diseases. This section will cover the identification, implications, and management strategies to keep your colony healthy.

4.4 Seasonal Hive Management

Understand how to adapt your beekeeping practices to the changing seasons. From preparing your hive for winter to maximizing productivity in the summer, this section will help you make timely adjustments throughout the year.

By the end of this chapter, you'll be equipped with the knowledge to effectively manage your hives, keeping your bees healthy, productive, and happy in any season. Let's buzz in and start mastering the art of hive management!

4.1 Routine Inspections: What to Look For

As we delve into the crucial task of routine hive inspections, remember that your role is similar to that of a bee colony's caretaker and detective. Regular checks are key to understanding the health and needs of your bees, catching issues early, and ensuring the colony thrives. Let's explore what you need to look out for during these inspections.

Visual Health Check of Bees: Start each inspection by observing the bees themselves. Healthy bees are active and purposeful in their movements. Watch for signs of disease or stress, such as discolored bodies, erratic behavior, or a noticeable decrease in activity. Healthy bees should be busy with tasks like foraging, feeding young, and constructing comb.

Brood Pattern Examination: The brood pattern tells a story about the queen's health and the hive's future. A strong, consistent brood pattern, with tightly packed larvae and capped cells, suggests a healthy queen. Look for the brood in a circular or oval pattern, which indicates good temperature control and hive health. Spotty patterns or missing brood could indicate problems with the queen or disease.

Queen Sighting: While you won't always see the queen during an inspection, knowing she's around is comforting. If you don't spot her, look for fresh eggs. A healthy queen lays eggs daily, and seeing new eggs assures you that the queen was active within the last three days.

Food Stores Check: Inspect the hive's stores of honey and pollen. These are essential for the bees' nutrition, especially heading into winter. Ensure there are adequate supplies; if stores are low, especially in early spring or late fall, you may need to feed your bees.

Comb Condition: Check the condition of the comb. It should be well-maintained and free from damage. Be on the lookout for mold, excessive propolis, or overcrowded comb, which can indicate ventilation issues or other problems in the hive.

Pest and Disease Signs: Always be vigilant for signs of pests and diseases. Common issues like Varroa mites, hive beetles, and foulbrood disease can devastate a hive if left unchecked. Look for mites on bees, larvae that fail to thrive, or irregular pupae caps, which might indicate brood disease. Regular monitoring helps you catch these problems early, significantly increasing your chances of managing them effectively.

Space and Congestion: Monitor the space within the hive. Overcrowding can lead to swarming, where the queen and many workers leave to find a new home. If the hive is too full, consider methods to give bees more space, like adding supers or managing the swarm impulse through splits.

Behavioral Observation: Finally, note the bees' behavior during the inspection. Aggressive or overly anxious bees might indicate underlying issues with the hive's health or disturbances in their environment.

Tools for the Task: Equip yourself with a good smoker, hive tool, and your protective gear. The smoker calms the bees, making your inspections safer and more productive. The hive tool is essential for gently prying apart frames and scraping off excess wax or propolis.

Conclusion and Record Keeping: After each inspection, record your findings. Keeping detailed records helps track the hive's history, monitor changes, and make informed decisions about interventions needed.

By conducting thorough and regular inspections, you're not just maintaining a hive; you're ensuring a thriving, productive community. This careful oversight is crucial to successful beekeeping and immensely rewarding as you watch your colonies grow strong and healthy. Happy inspecting, beekeeper!

4.2 Feeding Your Bees: When and How

Let's talk about a sweet and crucial aspect of hive management: feeding your bees. Just like any other living creature, bees need proper nutrition to thrive, especially when natural food sources are scarce. Understanding when and how to feed your bees can make a huge difference in their health and productivity. So, grab your sugar sacks and syrup mixers, and let's dive into the nectar-rich world of feeding bees!

Why Feed Your Bees? Bees naturally collect nectar and pollen from flowers, which they convert into honey—their food. However, there are times when flowers are scarce or weather conditions prevent bees from foraging. During such times, supplemental feeding is crucial to keep your colony alive and kicking. Common scenarios necessitating feeding include:

Post-Winter Recovery: After a long winter, bees may have exhausted their honey stores and need a boost to start the new season until natural nectar flows are available.

Newly Installed Colonies: New hives may lack sufficient stores and need feeding to help them get established.

Drought or Bad Weather: Unfavorable weather conditions can limit bees' ability to forage, requiring supplemental feeding to tide them over.

What to Feed Your Bees: The primary diet for bees you'll be mixing up is sugar syrup. The concentration of this syrup can vary depending on the season and the bees' needs:

1:1 Ratio (Sugar to Water): Used in spring and summer to stimulate comb building and support a growing colony.

2:1 Ratio (Sugar to Water): Used in fall to help bees build up their winter stores when they need more carbohydrates.

Pollen Substitutes: Sometimes, bees also need protein, especially in early spring or when pollen is scarce. Commercial pollen substitutes are available and can be provided via feeders inside or near the hive.

How to Feed Your Bees: Feeding bees is an art that requires gentle handling and the right equipment to ensure the bees get the nutrients without attracting pests or causing other problems.

Feeders:

Internal Feeders: These sit inside the hive, such as frame feeders, which replace one of the frames in a hive body. They're great for cooler weather as they don't expose bees to the elements.

External Feeders: Placed outside the hive, these can be jar feeders or entrance feeders. They're easier to refill but can attract robbers and pests.

Safety Tips: Always feed in the late afternoon or evening to reduce robbing. Ensure feeders are clean and free from mold. Check feeders regularly to replenish supplies and remove any dead bees.

Best Practices:

Avoid Overfeeding: Monitor your bees' intake. If they stop taking the feed, they might have found a natural source, or there could be issues within the hive that need addressing.

Hygiene: Keep the feeding area clean to prevent disease. Spoiled syrup or contaminated feeders can cause more harm than good.

Seasonal Adjustments: Tailor your feeding strategy to the season and the specific needs of your hive. What works in spring may not be suitable for fall.

Feeding your bees isn't just about keeping them alive; it's about ensuring they thrive. By providing the right type and amount of food at the correct times, you can support your bees through rough patches and help them build a strong, productive colony. So, keep your feeders filled, your syrup mixed, and your bees buzzing with health and happiness!

Remember, a well-fed bee is a happy bee, and happy bees make a happy beekeeper! Ready to tackle pests and diseases next? Let's keep the buzz going and ensure your hive remains a safe haven for your hardworking pollinators.

4.3 Managing Pests and Diseases

Just as gardeners must deal with weeds and pests, beekeepers must protect their hives from various threats. Pests and diseases can wreak havoc in a hive, but with careful management and a bit of bee-know-how, you can keep your colony healthy and buzzing. Let's dive into the strategies and best practices for managing the common pests and diseases that might target your precious pollinators.

Understanding the Threats: Before you can effectively protect your bees, it's crucial to know what you're up against. Some of the most common threats include:

- **Varroa Mites:** These tiny parasites can devastate a colony. They attach to bees and suck their hemolymph, weakening individual bees and spreading viruses.

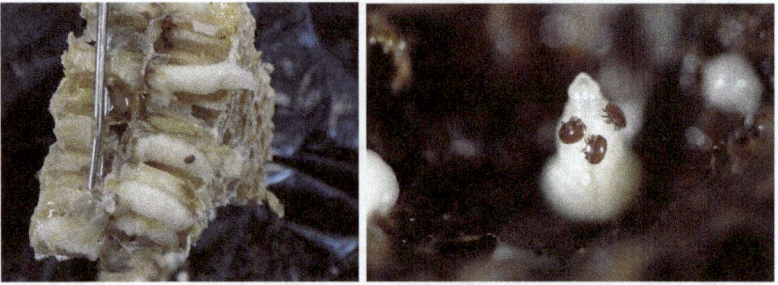

- **Nosema:** A microsporidian fungal infection that affects bees' digestive systems, leading to dysentery and weakening the colony's overall health.

- **American Foulbrood (AFB) & European Foulbrood (EFB):** These bacterial diseases can destroy larvae and are highly contagious. AFB spores can remain viable for years, making it particularly hard to eradicate.

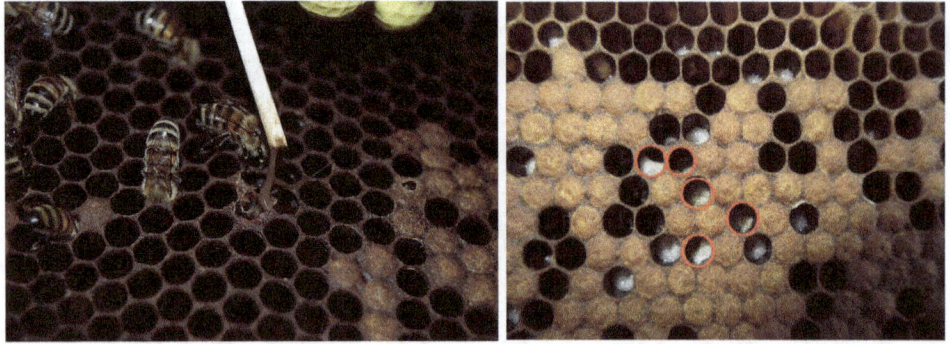

- **Small Hive Beetles and Wax Moths:** These pests can ruin comb, honey, and pollen stored within the hive.

Regular Inspections: The first line of defense is regular and thorough hive inspections. Look for signs of illness or infestation, such as:

- **Unusual bee behavior** like erratic movement or disorientation.
- **Physical signs on bees,** such as deformed wings or abdomens, which can indicate mite infestation.
- **Irregular brood patterns** or discolored larvae, which might suggest foulbrood or other diseases.
- **Damaged comb or hive structures**, often a sign of hive beetle or wax moth activity.

Preventative Measures: Prevention is always better than cure, especially in beekeeping where diseases and pests can spread rapidly.

- **Maintain Hive Hygiene:** Regularly clean and disinfect tools and equipment. Replace old combs with new ones every few years to prevent disease buildup.
- **Manage Hive Traffic:** Reduce the number of times you open the hive to minimize exposure to pests and stress to the bees.
- **Strong Colonies:** Ensure your bees are well-fed and healthy; strong colonies are better able to resist pests and diseases.

Treatment Strategies: When prevention isn't enough, it's time to treat. Here's how you can handle some common issues:

- **Varroa Mites:** Treat with miticides such as Apivar or natural treatments like oxalic or formic acid during times when honey supers are not present to avoid contamination of honey.
- **Nosema:** Provide affected colonies with fumagillin, an antibiotic that can help control the spread.
- **Foulbrood:** Infected colonies often need to be destroyed completely to prevent spread, as per local agricultural regulations.
- **Pests like Small Hive Beetles or Wax Moths:** Use physical traps inside the hive, and maintain strong colonies; healthy bees often keep these pests in check by themselves.

Monitoring and Record-Keeping: Keep detailed records of inspections, treatments, and any signs of disease or pest issues. Monitoring changes over time can help you identify patterns and understand the effectiveness of your management strategies.

By staying vigilant and proactive, you can manage pests and diseases effectively, ensuring your hives remain healthy and your bees continue their vital role in our ecosystem. Ready to adjust your management practices according to the seasons? Let's explore seasonal hive management next and keep our bees thriving all year round!

4.4 Seasonal Hive Management

Managing your bee colony throughout the year involves adapting your beekeeping practices to suit the changing seasons. Each season brings its own set of challenges and opportunities for bee health and productivity. Let's explore how to effectively manage your hives season by season, ensuring your bees stay busy and healthy all year round!

Spring Into Action: Spring is a time of growth and renewal, both for nature and your bee colony.

- **Inspection and Cleaning:** Start with a thorough spring inspection. Check for any winter damage to the hive and clean out debris. Look for signs of disease or pests that need to be addressed promptly.
- **Feeding:** Early spring can still have cold snaps that reduce natural foraging opportunities. Feeding your bees sugar syrup can give them the boost they need until flowers begin to bloom abundantly.
- **Swarm Prevention:** As the colony grows, watch for signs of swarming. Consider methods like splitting the hive to manage space and keep the colony size in check.

Summertime Buzz: Summer is peak season for bees, with long days and plenty of foraging available.

- **Hive Maintenance:** Regular inspections are crucial to ensure the hive is healthy and the queen is productive. Look for overcrowded conditions and add supers as needed for honey storage.
- **Water Supply:** Make sure there's an adequate water source nearby. Bees need water to cool the hive and process honey, especially on hot days.
- **Pest Management:** Keep an eye out for mites and other pests that can thrive in the warm weather. Treat appropriately to prevent infestations.

Autumn Preparations: As the weather cools, bees begin to prepare for winter, and so should you.

- **Harvesting Honey:** Late summer or early autumn is typically when you'll harvest honey, ensuring you leave enough for the bees to sustain themselves over winter.

- **Winter Preparations:** Begin to reduce the size of the hive by removing empty supers and consolidating the colony. Insulate the hive to protect against cold weather and ensure ventilation to prevent moisture buildup.
- **Feeding:** If natural food sources are scarce, feed your bees sugar syrup or fondant to build up their stores.

Winter Watchfulness: Winter is a quieter time for beekeepers, but there are still tasks to keep your colony secure.

- **Check Insulations:** Regularly check that the hive's insulation and wind barriers are intact.
- **Emergency Feeding:** Monitor the weight of the hive to gauge food stores. Feed fondant or dry sugar if stores are low, especially towards the end of winter.
- **Disease Checks:** Although inspections are minimal, keep an eye out for any signs of disease or distress during warm spells.

By understanding and implementing these seasonal management strategies, you can ensure that your bees remain vigorous and productive throughout the year. Whether you're prepping for spring blooms, summer harvests, autumn wind-down, or winter chill, your attentiveness to the hive's needs during each season is crucial to your success as a beekeeper. Ready to adapt and thrive? Let's keep our bees buzzing happily, whatever the weather!

Chapter 5: Understanding and Managing Swarms

In Chapter 5 we delve into one of the most dramatic behaviors exhibited by honeybees: swarming. Swarming is a natural part of a bee colony's life cycle, often misunderstood as aggressive behavior. However, it's actually a sign of a healthy and expanding colony. In this chapter, we'll explore the causes of swarming, learn strategies to prevent it, and discuss effective responses if your bees do decide to swarm.

5.1 What Causes Swarming?

Here, we'll uncover the biological and environmental triggers that lead a colony to swarm, such as overcrowding, an old or failing queen, and abundant food supplies.

5.2 How to Prevent Swarming

Preventing swarming is key to maintaining your bee colony within your hive. We'll explore practical steps you can take to manage hive space and colony health to minimize the chances of swarming.

5.3 What to Do If Your Bees Swarm

Despite your best efforts, swarms can still occur. This section provides a guide on how to safely capture and rehome a swarm, ensuring the safety of the bees and the public, and possibly even expanding your apiary in the process.

By the end of this chapter, you'll be well-prepared to understand, prevent, and manage swarming, keeping your colony thriving and productive. Let's buzz into the dynamic world of bee swarming and learn how to manage this fascinating aspect of bee behavior!

5.1 What Causes Swarming?

We're diving deep into one of the most natural yet complex behaviors of honeybees—swarming. While the sight of a bee swarm might strike awe or fear into the hearts of many, for beekeepers, it's a fascinating and critical aspect of bee colony dynamics. Swarming is essentially a colony's strategy for reproduction, growth, and survival. Let's explore the myriad factors that can trigger this captivating behavior.

Understanding Swarming: Swarming typically occurs when part of a bee colony leaves with the old queen to form a new colony. This usually happens in the spring, but understanding the precise triggers can help you manage or even prevent swarming to keep your colony strong and intact.

1. Overcrowding: One of the most common triggers for swarming is overcrowding within the hive. Bees are meticulous in their organization, and space is crucial for their operation. When a hive becomes too crowded, bees may feel stressed and begin preparations for swarming to alleviate the congestion. This is their way of finding more space and resources to continue their life cycle.

2. Aging Queen: The age and vitality of the queen can significantly influence swarming. An older queen, whose pheromone production has declined, may not be able to maintain the cohesion of the large colony. This reduction in pheromone levels can lead to reduced control over the colony, prompting bees to raise new queens and eventually leading some of the population to leave with the old queen.

3. Abundant Resources: Ironically, a very successful colony with abundant resources can also decide to swarm. When bees have plenty of food and the hive is thriving, they may take this as a cue that they have the resources to support a new colony. This abundance can trigger the colony to split, with part of the group venturing out to capitalize on their good fortune elsewhere.

4. Poor Ventilation and High Temperatures: Internal conditions of the hive, particularly ventilation and temperature, play a crucial role in swarming. Poor ventilation and high temperatures inside the hive can make it uncomfortable for the bees, pushing them to seek a new home with more suitable living conditions.

5. Genetics: Some bee strains are more prone to swarming than others. Genetics can influence the swarming behavior, with certain types more likely to swarm under similar conditions than others. Understanding the genetic propensity for swarming is important when selecting bees, especially if you aim to minimize this behavior.

6. External Factors: Various external factors such as nearby disturbances, regular hive manipulations by the beekeeper, or even the presence of pests and predators can stress the bees and contribute to swarming decisions.

Preventive Measures: Understanding these triggers is the first step in developing strategies to prevent unwanted swarming. Managing space within the hive, ensuring a healthy and youthful queen, maintaining optimal internal hive conditions, and minimizing stressful disturbances can help control swarming tendencies.

A Natural Phenomenon: It's important to remember that swarming is a natural and healthy behavior for bees. It is their way of ensuring survival and spreading their genetic material across the landscape. While beekeepers often wish to prevent swarming to maximize honey production and maintain colony strength, the impulse to swarm is a sign of a healthy bee colony.

In Summary: Swarming is driven by multiple complex factors that interplay in fascinating ways. By understanding these dynamics, you can better manage your hives and appreciate the incredible instincts of your bees. Whether you aim to prevent swarming or manage it effectively when it occurs, knowledge is your most powerful tool.

In the next section, we'll explore specific strategies on how to prevent swarming, ensuring your beekeeping efforts are as fruitful and rewarding as possible. Get ready to dive deeper into the hive mind and continue our journey into the buzzing world of bees!

5.2 How to Prevent Swarming

Swarming can be a significant event in the life of a bee colony, but managing it effectively is crucial to maintaining your hive's strength and maximizing honey production. Preventing swarming not only keeps your colony intact but also helps ensure a more stable and productive beekeeping experience. Let's buzz through the proactive measures you can take to minimize the likelihood of your bees deciding to pack up and move.

Understanding the Drive to Swarm: Before we get into the prevention techniques, it's essential to understand that swarming is a natural response to certain conditions within the hive. It's the colony's way of reproducing and ensuring survival. By addressing these triggers, we can significantly reduce the incidence of swarming.

1. Manage Hive Space Efficiently: Overcrowding is the most common reason bees decide to swarm. They need ample space not only to live but to store honey and pollen.

- **Regular Expansion:** Monitor the hive's capacity regularly and add extra supers or brood boxes before space becomes limited. This gives your bees room to grow and reduces the cramped conditions that can lead to swarming.
- **Honeycomb Management:** Removing old combs and replacing them with new frames can encourage the bees to stay and rebuild rather than leave. This also helps keep the hive clean and reduces disease.

2. Maintain a Young and Productive Queen: Aging queens produce fewer pheromones, which can decrease their ability to hold the colony together. Younger queens are more prolific and have a stronger pheromonal presence that can suppress the urge to swarm.
- **Requeening:** Consider replacing your queen every 1-2 years. This ensures robust pheromone levels and a vigorous, fertile queen that keeps laying plenty of eggs to keep the colony population stable.

3. Provide Optimal Conditions: Ensure that your hive environment promotes comfort and health for the bees, reducing stress factors that can trigger swarming.
- **Ventilation and Temperature Control:** Proper airflow helps regulate the temperature inside the hive, keeping it cool and comfortable for the bees.
- **Regular Cleaning:** Keep the hive free from debris and excess propolis build-up, which can obstruct the bees' movement and contribute to their discomfort.

4. Swarm Prevention Techniques: Implementing specific beekeeping techniques can also help prevent swarming:
- **Swarm Control Splits:** If your hive is growing rapidly, preemptively splitting the hive can help. By manually dividing the colony, you simulate the natural swarming process in a controlled manner, creating a new colony and reducing the population pressure in the original hive.
- **Checkerboarding:** This technique involves alternating empty frames with honey-filled frames in the upper boxes. It tricks the bees into thinking they have more space to fill and can delay or prevent swarming.

5. Monitor and Interact Regularly: Frequent checks allow you to spot early signs of swarming and take immediate action. These include the presence of queen cells (special cells built to raise new queens) which are a primary indicator that the colony is preparing to swarm.
- **Queen Cell Management:** Regularly inspect for and remove excess queen cells. This discourages the bees from raising new queens for swarming.

6. Feeding Appropriately: Incorrect feeding can contribute to swarming if not managed properly. Ensure that you feed your bees only when necessary, such as during scarce nectar flows, and avoid overfeeding, which can accelerate brood production and lead to overcrowding.

Preventing swarming is about understanding your bees' needs and behaviors and managing them proactively. By keeping your bees comfortable, well-spaced, and well-cared for, you reduce the chances of them feeling the need to leave the hive. With these strategies in place, your bees are more likely to stay put, stay productive, and keep those golden jars of honey coming.

Remember, every hive is unique, and what works best can vary depending on local conditions and the specific characteristics of your bee colony. Keep buzzing, keep learning, and keep loving those bees!

5.3 What to Do If Your Bees Swarm

Ah, the swarm—a beekeeper's bittersweet ballet. While it can be disheartening to see part of your colony take flight, swarming is a natural and healthy process for bees, signifying a strong and thriving colony. However, managing a swarm effectively is crucial to ensure the safety of the bees, the public, and possibly to reclaim or rehome the swarm. Here's your comprehensive guide on what to do if your bees decide to embark on this awe-inspiring venture.

Stay Calm and Observe: First things first, keep your cool. Swarming bees are generally not aggressive, as they have no hive to protect. They're also loaded with honey they consumed in preparation for their journey, making them less inclined to sting.

1. **Locate the Swarm:** Carefully observe where the bees are clustering. During a swarm, bees often gather around the queen on a nearby branch, fence, or other objects. This resting phase can last from a few hours to a couple of days as scout bees search for a new permanent home.

2. **Prepare Your Equipment:**
   - **Bee Suit:** Ensure you're fully suited in protective gear to avoid any stings.
   - **Bee Brush:** Have a gentle bee brush handy to coax any straggling bees.
   - **Hive Box:** Ready a new hive box close to the swarm site. This will house the swarm if you can successfully capture them.
   - **Ladder:** If the swarm is elevated, you'll need a ladder to reach them safely.

Capture the Swarm:

3. **Gently Gather the Bees:**
   - If the swarm is within reach, gently shake the branch or object they are clinging to, causing the cluster to fall into the open hive box placed below. If shaking is not feasible, you can gently brush the bees into the box.
   - Alternatively, if the swarm is on a structure that can't be shaken or brushed, place the open hive box directly underneath with a ramp leading into it. The bees often will move into the box on their own.

4. **Secure the Queen:** Ensuring that the queen gets into the box is crucial as the rest of the swarm will follow her. Once she's in, most of the other bees will move in after her. Occasionally check and gently tap the box to encourage stragglers.

5. **Provide a Warm Welcome:** Once the swarm is in the new hive, feed them with sugar syrup to keep them energized and encourage them to stay. Place frames with wax foundation in the box to promote comb building. Ensure the hive is well-ventilated and placed in a quiet location.

Aftercare:

6. **Monitor the New Hive:** Keep a close eye on the new setup. Ensure that the bees are adapting to their new environment, starting to build comb, and that the queen is laying eggs.

7. **Check for Diseases:** As you didn't have control over where the swarm came from, it's vital to monitor for any signs of diseases or pests that could spread to your other hives.

8. **Integration or Relocation:** Decide whether you will integrate the new swarm into your existing apiary or relocate them. If integrating, ensure that the new hive does not compete for resources with your existing colonies.

Educate and Engage: If the swarm happens in a public area, use this as an opportunity to educate onlookers about bee behavior, the importance of pollinators, and how swarming is a natural occurrence. This can help ease fears and spread awareness.

Legal and Ethical Considerations: Always ensure you are complying with local wildlife and beekeeping laws when capturing and relocating swarms. Respect property boundaries and seek permissions if necessary. Swarming can initially seem like a crisis, but with the right knowledge and tools, it can be a golden opportunity to expand your apiary and contribute to the health and growth of your beekeeping endeavors. Embrace it as a part of the natural cycle of bee life and a testament to your success as a beekeeper. Now that you know what to do when your bees swarm, you're better prepared to handle this buzzing event with confidence and care!

Chapter 6: Harvesting Honey

Welcome to Chapter 6, where the sweet rewards of your beekeeping efforts come to fruition—harvesting honey! This chapter will guide you through the exciting process from deciding the right time to harvest, to safely extracting honey, processing it, and finally storing it for your enjoyment. Additionally, we'll explore other valuable hive products such as wax, propolis, and royal jelly. Each section is designed to help you maximize your harvest while ensuring the health and sustainability of your bee colony.

6.1 When to Harvest Honey

Learn how to identify the perfect timing for honey collection, which depends on your local climate, the condition of the hive, and the type of plants available to your bees. Timing is crucial to ensure that the honey is at its highest quality and that you don't deplete your bees' essential reserves.

6.2 How to Harvest Honey Safely

Safety is key in honey harvesting, not just for you but for your bees. This section will cover techniques and tips to extract honey with minimal disruption and stress to the colony. You'll learn about using smokers, wearing appropriate gear, and the best practices for removing frames filled with honey.

6.3 Processing and Storing Your Honey

Once harvested, honey needs to be processed and stored properly to preserve its natural flavors and beneficial properties. We'll go through filtering, bottling, and the ideal conditions for storing honey to maintain its quality over time.

6.4 Other Hive Products: Wax, Propolis, and Royal Jelly

Beyond honey, your hive is a treasure trove of other valuable products. This section will delve into how to harvest, process, and use beeswax, propolis, and royal jelly—each with unique benefits and uses in both daily life and various health and wellness applications.

By the end of this chapter, you'll be equipped with all the knowledge you need to effectively harvest and make the most of your hive's products. Let's get started on this sweet journey to harvesting and beyond!

6.1 When to Harvest Honey

As we dive into the sweetest chapter of your beekeeping journey, it's essential to know not just how to harvest honey, but when. Timing your honey harvest is crucial for both the health of your bees and the quality of the honey you gather. So, let's explore the buzzing world of bees and their liquid gold, ensuring you're equipped to choose the perfect moment to harvest.

Understanding the Bee Lifecycle and Honey Production: Honey is the beautiful result of bees collecting nectar from flowers. This nectar then undergoes a magical transformation within the hive, as bees add enzymes that convert its sugars. The bees fan the nectar with their wings, helping the water evaporate and leaving behind thick, sweet honey. Knowing this process helps you understand why timing is so crucial.

Seasonal Considerations: The timing of honey production and optimal harvest periods can vary significantly depending on your local climate and the floral sources available to your bees.

- **Spring:** In many regions, spring can provide a flush of floral nectar from blooming plants and trees. It's a prime time for bees to start building up their honey stores.
- **Summer:** This season often marks the peak of honey production. Flowers are abundant, and long days allow bees more time to forage. However, as summer progresses, nectar flow can diminish, especially in areas with hot, dry summers.
- **Fall:** Some areas enjoy a late flow of nectar from plants that bloom in late summer or early autumn. However, in many climates, fall is the time to start preparing bees for winter rather than expecting significant honey production.

Signs That Honey is Ready for Harvest: Bees will let you know when it's time to harvest by capping the honeycomb cells with wax. Here's what to look for:

- **Capped Honey:** A frame ready for harvesting will be about 80% capped. This means the bees have sealed off the honey cells with wax, indicating that the honey is dried to their standards and preserved for long-term storage.
- **Frame Inspection:** Gently lift the frames to inspect both sides. If you see that most of the comb is capped, it's a good indicator that it's time to harvest.
- **Checking Honey Moisture Content:** If you want to be extra sure about the quality of the honey, you can use a honey refractometer to check the moisture content. Honey should generally be below 18% moisture to prevent fermentation.

Impact on the Colony: Harvesting honey must be done thoughtfully to ensure that your bees have enough stores to get them through the winter.

- **Leave Enough Honey for Bees:** Always ensure there is sufficient honey left in the hive for the bees. A good rule of thumb is to leave at least 15 to 20 pounds of honey in the hive for the bees, depending on your local climate and the length of your winter.
- **Feeding After Harvest:** If necessary, consider feeding your bees sugar syrup in the autumn to help them build up their winter stores if their natural honey reserves are insufficient.

Ethical and Sustainable Harvesting Practices: Harvesting honey responsibly ensures the sustainability of your beekeeping practice.

- **Avoid Over-Harvesting:** Taking too much honey can stress the colony, potentially leading to health problems or even colony collapse.

- **Monitoring Bee Health:** Keep an eye on your bees' health throughout the harvesting process. If the colony appears weak or stressed, reduce the amount of honey taken or adjust your management practices.

By understanding these principles, you're ready to make informed decisions about when to harvest honey, balancing the needs of your bees with the joy of gathering your sweet rewards. Remember, good things come to those who wait—the patient beekeeper often reaps the richest, most delicious honey. So, tune in to your bees, and let them guide you to the perfect harvest time!

6.2 How to Harvest Honey Safely

Hello, conscientious beekeepers! Harvesting honey is undoubtedly one of the most rewarding aspects of beekeeping, offering the sweet fruits of your labor and your bees'. However, it's crucial to approach this process with care to ensure safety for both you and your bees. Let's explore the best practices for harvesting honey safely, maintaining the integrity of the hive, and ensuring that your bees remain healthy and stress-free.

Preparing for Harvest: Before you dive into harvesting, preparation is key. Here's how you can set the stage for a smooth and safe honey collection:

1. **Choose the Right Time:** Early to mid-morning on a sunny day is ideal. Bees are more likely to be out foraging, which means fewer bees at the hive and less stress for both you and the bees during the harvest.
2. **Wear Protective Gear:** Always wear a full bee suit, including gloves and a veil. This protective clothing isn't just to prevent stings; it also keeps you calm and confident, knowing you're protected as you work.
3. **Have the Right Tools:**
    - **Bee Brush:** Gently brushes bees off the combs without harming them.

- **Hive Tool:** Helps in prying apart hive components and scraping wax.
- **Smoker:** A smoker can calm the bees and make them less aggressive by masking their pheromones.

- **Uncapping Knife or Fork:** To remove the wax caps from the honeycombs.

- **Extractor:** A device that spins the frames, forcing honey out of the comb by centrifugal force.

Harvesting Process: Follow these detailed steps to safely remove honey from your hive:

4. **Smoke the Hive:** Lightly smoke the hive entrance and under the lid to calm the bees, which makes them less likely to become defensive.
5. **Remove the Supers:** Carefully remove the supers (honey storage boxes) from the hive. Ensure you have a clean, bee-free area where you can work on extracting the honey.
6. **Brush Off Bees:** Gently brush off any remaining bees from the frames using the bee brush. It's important to do this calmly and gently to minimize stress on the bees.
7. **Uncap the Frames:** Use an uncapping fork or knife to scrape off the wax caps from the honeycomb. This exposes the honey for extraction.
8. **Use the Extractor:** Place the uncapped frames into the honey extractor. The extractor spins the frames, using centrifugal force to pull the honey out of the comb.
9. **Strain and Collect Honey:** As honey is extracted, it should be strained through a mesh filter to remove any wax pieces and impurities. Then, collect the honey in a clean bucket or directly into storage containers.
10. **Return Frames to the Hive:** After extraction, return the wet frames to the hive. Bees will clean up the remaining honey and repair the comb, which not only saves their energy but also prepares the frames for future honey production.

Post-Harvest Tips: After harvesting, consider these additional tips to ensure ongoing hive health and productivity:

11. **Check Hive Health:** Post-harvest is a good time to check for signs of disease or distress in the hive. Ensure your bees are healthy and the queen is actively laying.
12. **Manage Hive Resources:** Ensure that your bees have enough honey left for their needs, especially heading into winter. If stores are low, you may need to feed your bees.
13. **Clean Your Equipment:** Thoroughly clean all your tools and equipment with hot, soapy water to prevent disease spread and to be ready for the next use.

Harvesting honey safely is all about preparation, patience, and respect for your bees. By following these guidelines, you can enjoy the sweetness of your harvest without causing undue harm or stress to your bee colony. Here's to a bountiful and safe honey harvest!

6.3 Processing and Storing Your Honey

After the thrilling experience of harvesting, we come to a crucial stage: processing and storing your honey to ensure it maintains its purity, flavor, and nutritional value. Proper processing and storage are key to preserving the delightful essence of your hard-earned honey. Let's explore the best practices to handle your honey from hive to table, ensuring it remains perfectly delightful.

Processing Honey: Once you've extracted honey from the combs, it's not quite ready to enjoy or sell. Here's how to refine it further:

1. **Filtering:**
   - **Purpose:** To remove any remaining wax particles, bee parts, or debris collected during extraction.
   - **Method:** Use a fine mesh filter or a nylon straining bag. Avoid using metal as it can impart undesirable flavors to the honey.

   - **Tip:** Don't filter it too finely if you wish to keep valuable pollen in the honey, which is a selling point for many local honey enthusiasts.

2. **Settling:**
   - **Purpose:** Allows tiny air bubbles and fine particles to rise to the top after filtering.
   - **Duration:** Let the honey sit in a settling tank or a clean food-grade bucket for a few days in a cool, dark place.
   - **Skimming:** After settling, skim off any foam and particles that have accumulated at the top.

3. **Decanting:**
   - **Purpose:** To separate the pure honey from any residue or foam that did not get removed during skimming.

- **Process:** Carefully pour or pump the clear honey into storage containers, avoiding disturbing the bottom layer of the settling tank.

Storing Honey: Proper storage is vital to preserving the quality and extending the shelf life of your honey.

4. **Container Selection:**
   - **Materials:** Use food-grade containers made of glass, stainless steel, or BPA-free plastic. Glass is ideal for long-term storage as it doesn't impart any flavors and allows visibility of the honey's clarity.
   - **Sizes:** Depending on your distribution plans, you might choose large drums or small retail-ready jars.

5. **Temperature and Environment:**
   - **Cool and Dark:** Store honey in a cool, dark place away from direct sunlight. Sunlight can degrade the quality of honey over time.
   - **Stable Temperature:** Fluctuating temperatures can lead to honey crystallization. While crystallization doesn't affect the quality, it can alter the texture, making it grainy.

6. **Avoid Moisture:**
   - **Seal Well:** Ensure all containers are tightly sealed. Moisture can lead to fermentation, which spoils the honey.
   - **Handling Care:** Keep moisture out by avoiding leaving containers open and using dry, clean utensils when handling honey.

7. **Long-Term Considerations:**
   - **Crystallization:** Over time, most natural honey will crystallize. You can gently re-liquify it by warming the jar in a hot water bath, ensuring the temperature does not exceed 40 degrees Celsius (104 degrees Fahrenheit) to preserve natural enzymes.

8. **Labeling:**
   - **Information:** Properly label your honey jars with the harvest date, origin, and any specific floral notes if known. This information is particularly cherished by honey connoisseurs and helps in tracking batches.

By adhering to these detailed processing and storage guidelines, you ensure that the honey you've worked so hard to produce retains its highest quality, flavor, and nutritional benefits. Whether you're preparing to sell, gift, or enjoy your honey, these steps will help you keep your honey as delightful as the day it was harvested. Let's keep our honey sweet and our beekeeping sweeter!

6.4 Other Hive Products: Wax, Propolis, and Royal Jelly

While honey might be the sweet spotlight of beekeeping, there are several other valuable products that your hive produces. Beeswax, propolis, and royal jelly are treasures in their own right, each boasting unique properties and uses that can enhance your beekeeping rewards. Let's dive deep into the world of these fascinating hive products, exploring how to harvest, process, and utilize them effectively.

Beeswax: Nature's Craft Material Beeswax is a byproduct of the honey production process, secreted by bees to build the hexagonal cells in which honey is stored and larvae are grown.

- **Harvesting Beeswax:**
    - Collect the wax caps that you slice off during the honey extraction process.
    - Gather old or damaged combs that are no longer suitable for the bees.
- **Processing Beeswax:**
    - Melt the wax using a solar wax melter or a double boiler method, ensuring not to overheat and degrade the wax.
    - Filter the melted wax through a fine mesh to remove impurities.
- **Uses of Beeswax:**
    - Craft beautiful candles, soaps, and cosmetics.
    - Use in woodworking and sewing as a lubricant and waterproofing agent.
    - Produce food wraps as a natural alternative to plastic cling films.

Propolis: The Bee's Defense Propolis is a resin-like material made by bees from substances collected from trees and other botanical sources. It is used within the hive to seal small gaps and protect the colony from infections.

- **Harvesting Propolis:**
    - Install a propolis trap, which is a flexible sheet with small holes placed at the top of the hive; bees fill these holes with propolis, which can then be frozen and flexed to harvest the propolis.
    - Scrape excess propolis from hive frames and boxes during inspections.
- **Processing Propolis:**
    - Freeze the propolis to make it brittle for easier handling and cleaning.
    - Dissolve clean propolis in alcohol to make a tincture or in oil for other applications.
- **Uses of Propolis:**
    - Natural health remedies, particularly for its antibacterial and antifungal properties.
    - Dental care products, as an anti-inflammatory and antimicrobial agent.
    - Wood varnishes and sealants, given its durable and protective properties.

Royal Jelly: A Superfood from the Hive Royal jelly is a creamy secretion produced by young nurse bees, primarily to feed the queen and young larvae. It is highly nutritious and considered a superfood in many cultures.

- **Harvesting Royal Jelly:**
  - Requires careful timing and involves using queen cups to encourage nurse bees to produce royal jelly.
  - Extract royal jelly from the queen cells a few days after the larvae have been fed with it but before they start to transform the food into bee tissue.
- **Processing Royal Jelly:**
  - Collect it using a small spatula or syringe.
  - Keep it refrigerated or freeze it immediately to preserve its freshness and nutritional properties.
- **Uses of Royal Jelly:**
  - Dietary supplements, celebrated for enhancing immunity and energy.
  - Skin care products, due to its anti-aging, moisturizing, and healing properties.
  - In culinary applications, albeit less common, for its health benefits.

Integrating Hive Products into Your Practice To fully embrace the potential of these hive products, consider incorporating them into your beekeeping routine:

- **Educate Yourself:** Learn more about the properties and handling of each product to maximize their use and benefits.
- **Market Wisely:** If you plan to sell these products, educate your customers about their unique benefits and uses.
- **Sustainability:** Always harvest these products sustainably, ensuring that your actions do not harm the hive or the bees' well-being.

By valuing all products that your hive produces, you elevate your beekeeping practice from simply harvesting honey to fostering a sustainable, multifaceted apicultural operation. Enjoy exploring these wonderful gifts from your bees, and remember, every part of the hive holds potential!

Chapter 7: Overwintering Your Bees

In this chapter we will discuss fundamental aspects of preparing your bees for the colder months. Overwintering your bees is vital for their survival and your success as a beekeeper. This chapter will guide you through the essential steps needed to ensure your hive thrives through the winter and emerges strong in the spring. We'll cover everything from preparing your hive for the cold, to specific feeding strategies, and how to tackle common winter challenges.

7.1 Preparing Your Hive for Winter:

Learn how to properly insulate your hive, manage ventilation, and make necessary adjustments to the hive's physical setup to help your bees conserve heat and stay dry during the cold months.

7.2 Winter Feeding and Care:

Discover the best practices for feeding your bees when natural food sources are scarce, including what to feed, how much, and how to administer this sustenance without exposing the hive to harsh elements.

7.3 Common Winter Challenges:

Understand the typical challenges that beekeepers face during the winter, such as colony starvation, moisture control, and pest management, and learn effective strategies to overcome these issues.

By the end of this chapter, you'll be equipped with the knowledge to help your bees not just survive but thrive during the winter, setting the stage for a productive spring season. Let's get your bees snug and secure for the winter!

7.1 Preparing Your Hive for Winter

As the crisp air of autumn begins to bite, it's time to think about preparing your hives for the colder months ahead. Winter can be a challenging time for bees, and the steps you take now to secure their comfort and safety can have a significant impact on their survival and spring vigor. Let's walk through a comprehensive guide on how to prepare your hive for winter, ensuring your bees are cozy, well-fed, and ready to face the cold.

Understanding the Bee's Winter Needs: Bees are remarkably adept at managing their internal hive temperature during winter. They cluster together, vibrating to generate heat, and maintaining an impressively warm core temperature to keep the queen and themselves alive. However, they need your help to create the best conditions for this process.

1. Hive Inspection and Maintenance: Before the cold sets in, a thorough inspection and maintenance routine is crucial:

- **Check Hive Health:** Ensure your hive is free from diseases and pest infestations like Varroa mites, which can weaken the bees during winter. Treat any issues before winter fully sets in.
- **Repair Cracks and Gaps:** Seal cracks and gaps in the hive body to prevent drafts that can chill the bees. However, ensure you do not seal off all ventilation as moisture build-up is another significant winter risk.
- **Reduce Hive Space:** Consider reducing the space the bees need to keep warm. Use a hive reducer or rearrange the frames to help them conserve heat more efficiently.

2. Ventilation Management: Proper ventilation is crucial to prevent moisture from accumulating inside the hive, which can be more deadly than the cold itself:

- **Upper Ventilation:** Ensure there is adequate ventilation at the top of the hive to allow warm, moist air to escape, preventing condensation from dripping back onto the cluster.
- **Avoid Over-Ventilation:** While ventilation is necessary, too much can lead to excessive cold air flow, which can chill the bees. Striking the right balance is key.

3. Insulation Techniques: Insulating your hive helps the bees maintain their required warmth with less energy expenditure:

- **Wrap the Hive:** Use commercially available hive wraps or natural insulating materials like burlap sacks filled with straw. Wrap the hive snugly, but ensure the entrance and ventilation points are not obstructed.
- **Positioning:** Consider the hive's exposure to elements. Position it to receive maximum sunlight during the day. Shield it from prevailing winter winds, possibly by relocating the hive to a more sheltered spot or setting up a windbreak.

4. Food Stores Check: Ensuring your bees have enough food to last through winter is possibly the most crucial aspect of winter preparation:

- **Honey Stores:** Bees need approximately 30 to 60 pounds of honey stored away to make it through winter, depending on your local climate.
- **Supplemental Feeding:** If honey stores are insufficient, provide emergency feeding with sugar cakes or fondant. Place these directly above the cluster where bees can access them without having to travel far in the cold.

5. Pest and Rodent Control: Rodents seeking warmth might see your hive as a winter refuge, which can be disastrous:

- **Mouse Guards:** Install mouse guards at the entrance of the hive. These prevent mice from entering but allow bees to pass through.

- **Check Surroundings:** Regularly check and clean the area around the hive to reduce the attraction for rodents.

6. **Continuous Monitoring:** Even after all preparations, regular winter checks are essential:
   - **Brief Inspections:** On warmer winter days, take the opportunity to quickly check on the health of the cluster and the food supply. Avoid opening the hive too much, as this can release much-needed heat.

7. **Record Keeping:** Maintain records of your winter preparations and observations. This information will be invaluable for learning and improving over subsequent winters.

By taking these detailed steps to prepare your hive for winter, you're setting up your bees for a successful overwintering. The work you do now supports not just their survival but also their ability to thrive and grow when spring arrives. Let's keep our bees warm, well-fed, and ready for a new season!

7.2 Winter Feeding and Care

Winter can be a harsh time for your bees, but with proper feeding and care, you can help ensure they stay healthy and strong until the spring flowers bloom. Feeding your bees during the colder months isn't just about keeping them alive; it's about maintaining their health and vitality so they can hit the ground running when warmer weather returns. Let's explore how to effectively feed and care for your bees during winter.

Understanding the Importance of Winter Feeding: During winter, bees cluster together and use their stored honey as their primary food source to generate heat and maintain a viable temperature within the hive. However, if their honey reserves are low—due to a less productive summer or higher consumption needs—they will require supplementary feeding to survive the winter.

Assessing the Hive's Food Stores: Before the onset of cold weather, you should perform a thorough check of the hive's honey reserves. A healthy colony typically needs about 30 to 60 pounds of honey to get through the winter, depending on the severity of your climate. If the frames are not well-capped with honey, you'll need to prepare supplementary food.

Types of Winter Feed:

1. **Sugar Syrup:** Early winter or in regions with milder winters, bees can still consume liquid feed. Heavy syrup (2:1 ratio of sugar to water) is preferred as it provides high energy content and is less prone to fermentation.
2. **Fondant or Candy Boards:** As temperatures drop further, liquid feeding can cause too much moisture inside the hive. Fondant or sugar candy boards provide a drier alternative, supplying essential carbohydrates without increasing hive humidity.

3. **Dry Sugar Feeding:** An emergency feeding method involves providing plain granulated sugar, which the bees use to draw moisture and create a consumable syrup. This method can be particularly useful if the winter turns out harsher than expected and other food stores are exhausted.

How to Feed Your Bees in Winter:
- **Placement:** Place food directly above the cluster where bees can access it without needing to break cluster. This placement is crucial because bees will not venture far from the cluster in cold weather.
- **Insulation:** When adding feed, especially in the form of fondant or candy boards, ensure that the hive top is well insulated. This setup helps maintain the warmth generated by the bees, making it easier for them to access the food.

Monitoring and Adjustments:
- **Regular Checks:** On warmer days, quickly check the food levels without fully opening the hive and releasing precious heat. Refill as necessary.
- **Moisture Control:** Too much moisture is a major winter hazard that can lead to mold and cold stress. Ensure your hive has adequate ventilation to allow moisture to escape, particularly when using sugar or syrup feedings.
- **Pest Management:** Keep an eye out for pests like mice, which can invade hives to eat the bees' food and destroy their comb. Ensure mouse guards are in place and properly secured.

Special Considerations for Weak Colonies:
- **Extra Care:** Weak or recovering colonies might have higher food needs. Be more vigilant with these hives and consider additional insulation or windbreaks to help them conserve energy.
- **Health Checks:** Monitor these colonies for signs of disease or distress, as they can be more susceptible during the stressful winter months.

Winter feeding is more than just providing food; it's about supporting the overall health of your hive in the harsh months. Proper preparation, regular monitoring, and understanding the unique needs of your bees during winter can ensure that they not only survive but thrive and are ready for the spring. Remember, a well-fed and well-cared-for colony is a productive colony when the warm weather returns. Let's keep our bees healthy and buzzing all winter long!

7.3 Common Winter Challenges

Winter presents a unique set of challenges that can test the resilience of your hive. Understanding these challenges is crucial for keeping your bees healthy and active until the spring thaw. Let's explore the common winter hurdles every beekeeper might face and arm you with strategies to overcome them effectively.

1. Starvation Risk: Perhaps the most critical challenge during winter is the risk of starvation. As bees cluster and temperatures drop, their movement slows, and their ability to access stored food can be compromised.

- **Preventive Measures:** Ensure your bees have ample honey stores before winter sets in. A typical colony needs between 30 to 60 pounds of honey to survive the winter, depending on the climate.
- **Mid-Winter Checks:** On milder winter days, quickly check the weight of the hive by gently lifting it from the back. If it feels light, consider emergency feeding options like fondant or dry sugar directly above the cluster.

2. Excessive Moisture: Moisture in the hive can be more deadly than the cold. Bees generate a significant amount of moisture through respiration, and if this moisture condenses inside the hive, it can drip onto the cluster, chilling and potentially killing the bees.

- **Proper Ventilation:** Ensure your hive has adequate ventilation at the top to allow moist, warm air to escape. This might involve adding small upper entrance holes or ensuring the inner cover allows for air movement.
- **Absorbent Materials:** Some beekeepers place absorbent materials such as wood shavings or newspaper above the inner cover to soak up excess moisture.

3. Cold Winds: Harsh winds can penetrate hive cracks and crevices, chilling the bees and forcing them to consume more honey to maintain their temperature.

- **Windbreaks:** Position your hives behind natural windbreaks like bushes or trees, or construct artificial ones such as bales of straw or a wooden barrier.
- **Hive Insulation:** Wrap hives with insulation wraps or use hive cozies to keep the bees warm. Be cautious not to block any crucial ventilation points.

4. Pest Infestations: Winter doesn't always stop pests. Mice and other rodents may seek shelter in the warm confines of your hive, and Varroa mites can continue to afflict the colony.

- **Pest Management:** Install mouse guards at the hive entrance to prevent rodents from entering. Regularly treat for Varroa mites during the late fall to reduce infestation levels going into winter.
- **Regular Monitoring:** Check for signs of pests during mild weather checks and address any infestations immediately.

5. Reduced Bee Population: The natural lifecycle of bees leads to a reduced population during winter. This smaller workforce means less heat generation and increased vulnerability.

- **Colony Strength Assessment:** In late summer and early fall, assess the strength of your colony. Weak colonies can be combined with stronger ones to ensure sufficient numbers to maintain warmth.

6. Queen Health: The health of the queen bee is essential to the spring rebound of the colony. A failing queen can result in a weak or non-existent brood in the spring.

- **Fall Assessment:** Evaluate the queen's performance and the brood pattern during the fall. Consider replacing a poorly performing queen before winter.

7. Spring Buildup Readiness: Post-winter, the colony must quickly build up to take advantage of early spring nectar flows. Being unprepared can delay this and affect the colony's productivity.

- **Early Spring Feeding:** As soon as temperatures allow, begin feeding a 1:1 sugar syrup to stimulate brood rearing and give your bees a head start.

By anticipating these challenges and implementing strategic measures, you can safeguard your bees against the harsh realities of winter. Each step you take enhances your hive's resilience, ensuring that when spring arrives, your colony is healthy, vibrant, and ready to buzz into the new season. With careful planning and a bit of winter wisdom, your bees will not just survive the cold months but thrive.

Chapter 8: Advanced Topics in Beekeeping

In this chapter where we delve into more sophisticated aspects of beekeeping for those looking to expand their expertise and explore new avenues within this fascinating field. This chapter is designed for seasoned beekeepers aiming to refine their skills and knowledge, covering everything from breeding queens to advanced disease management and venturing into bee-related products like mead making. Let's explore these advanced topics, which can enhance your beekeeping experience and offer new challenges and rewards.

8.1 Breeding Your Own Queens: Learn the art and science of queen rearing, which is crucial for maintaining strong, healthy colonies and can be a sustainable way to expand your apiary or provide valuable bees to other beekeepers.

8.2 Advanced Disease Management: Dive deeper into the strategies for diagnosing and treating bee diseases and pests. This section will equip you with the knowledge to handle complex health issues that can threaten your hives.

8.3 Making Mead and Other Bee Products: Explore the exciting world of mead making and other bee-derived products such as beeswax and propolis-based items. Discover how to creatively use every part of your hive's output to produce unique, marketable products.

Each section of this chapter is designed to build on your existing beekeeping knowledge, pushing you into new territories and helping you to become a master of the craft.

8.1 Breeding Your Own Queens

Welcome, ambitious beekeepers, to the intricate art of queen breeding—a skill that elevates your beekeeping experience and significantly enhances the vitality of your apiary. Breeding your own queens allows for greater control over your colonies' genetics, health, and productivity. This comprehensive guide will delve into the processes, benefits, and techniques needed to successfully breed and introduce new queens into your hives.

Understanding Queen Breeding: Queen breeding involves selecting desirable traits and systematically raising queen bees to take over existing colonies or to create new ones. This practice is essential for maintaining strong, healthy bee populations and can help address issues like disease resistance, temperament, productivity, and adaptability to specific climates.

Why Breed Your Own Queens?

1. **Genetic Diversity:** Breeding queens allows you to introduce new genetics into your hives, which can enhance traits like disease resistance, honey production, and gentleness.
2. **Colony Expansion:** With your own queens, you can expand your apiary sustainably without needing to purchase new bees.

3. **Replacement and Recovery:** Quickly replace underperforming or deceased queens, ensuring minimal disruption to your hive's productivity.

The Breeding Process: Breeding quality queens is a multi-step process that requires careful planning and execution:

- **Step 1: Selection of Breeder Queens:** Choose strong, healthy colonies with desirable traits as breeders. The genetic quality of the mother queen critically impacts the next generation.
- **Step 2: Drone Management:** Since drones (male bees) provide the genetic counterpart to the queen, managing drone populations is crucial. Encourage strong drone populations in your apiary to ensure diverse and robust mating.
- **Step 3: Grafting Larvae:** This involves transferring very young larvae (no older than 24 hours) from the breeder queen's comb into artificial queen cells. These cells are then placed in a queenless hive or a specially prepared "cell starter" colony to be raised as queens.
- **Step 4: Queen Rearing:** Once the larvae are grafted, the nurse bees in the starter hive feed and care for the developing queens. The environment must be meticulously managed to ensure optimal temperature and humidity.
- **Step 5: Mating Flights:** Once matured, virgin queens are placed in mating nucs where they will take their mating flights. This phase is critical as the queens mate with drones in flight, which determines the genetic quality of her offspring.
- **Step 6: Evaluation and Introduction:** After mating, each queen's performance is evaluated based on her egg-laying patterns and brood viability. Successful queens can then be introduced to new or existing hives that require a queen.

Tools and Equipment Needed:

- **Grafting tools:** Used for transferring larvae to queen cups.
- **Queen cups:** Artificial cells where queen larvae are reared.
- **Incubators:** For controlling climate during queen development.
- **Mating nucs:** Miniature hives used to house and mate new queens.

Challenges and Tips:

- **Weather Dependence:** Queen rearing is highly dependent on good weather for mating flights—poor weather can result in poorly mated queens.
- **Disease Management:** Maintain strict hygiene and monitor for diseases, as young queens are particularly vulnerable.
- **Timing:** Timing is crucial in queen breeding. Mistakes in timing can lead to the loss of larvae or failure of queens to properly mate.

Breeding your own queens is both a science and an art. It requires patience, precision, and a deep understanding of bee biology. However, the rewards of developing a resilient and tailored bee population are immense. By mastering queen breeding, you not only ensure the sustainability of your apiary but also contribute to the broader health and diversity of bee populations. Dive into this advanced aspect of beekeeping, and watch your colonies flourish under the leadership of queens you've carefully cultivated!

8.2 Advanced Disease Management

8.2 Advanced Disease Management

Welcome to a crucial aspect of beekeeping that challenges even the most experienced beekeepers: advanced disease management. In this segment, we dive deep into the strategies, practices, and innovations that help you safeguard your colonies from diseases and pests, ensuring their longevity and productivity. Understanding and managing bee health at an advanced level not only supports your apiary but also contributes to the broader beekeeping community by promoting healthier bees.

Understanding Bee Diseases and Pests: The first step in advanced disease management is a thorough understanding of the pathogens and pests that can afflict bees, including bacteria, viruses, fungi, and mites. Each of these has its own implications and requires specific approaches for control and treatment.

1. **Bacterial Diseases:** Such as American Foulbrood (AFB) and European Foulbrood (EFB), which affect the brood with different levels of severity. AFB, for example, is highly contagious and can lead to the destruction of infected colonies.
2. **Viral Diseases:** Like Deformed Wing Virus (DWV) and Israeli Acute Paralysis Virus (IAPV), often exacerbated by Varroa mite infestations, affecting bee development and colony vigor.
3. **Fungal Infections:** Nosema is a common fungal disease that affects bee digestion and can lead to weakened colonies, especially under stressful conditions.
4. **Parasites and Pests:** The Varroa mite is perhaps the most devastating, weakening bees by feeding on their bodily fluids and spreading diseases.

Advanced Diagnostic Techniques: Modern beekeeping now benefits from advanced diagnostic techniques that allow for early detection and precise identification of pathogens, which is crucial for effective management.

- **Lab Testing:** Regular submission of bee samples to laboratories can help detect diseases before they manifest visibly in the colony.
- **On-Site Kits:** Emerging technologies now offer on-site testing kits for rapid health assessment in the apiary.

Integrated Pest Management (IPM): IPM is a holistic approach to disease and pest control that combines different management strategies to minimize the use of chemicals, focusing instead on preventative measures and sustainable solutions.

- **Cultural Controls:** These include practices such as maintaining strong and genetically diverse colonies, which are less susceptible to disease and stress.
- **Mechanical Controls:** Techniques like drone brood removal and screen bottom boards help physically remove or reduce pest populations.
- **Biological Controls:** The use of natural predators or competitors to control pest populations is being explored in some areas.
- **Chemical Controls:** When necessary, the selective use of miticides and antibiotics, applied judiciously to minimize resistance build-up and contamination.

Enhanced Bee Nutrition: Supporting bee health through enhanced nutrition can significantly improve disease resistance and colony resilience.

- **Supplemental Feeding:** Providing balanced nutritional supplements during times of scarce natural forage can boost immunity and vitality.
- **Probiotics:** Emerging research into bee gut health suggests that probiotics can play a role in enhancing disease resistance.

Regular Monitoring and Record-Keeping: Vigilant monitoring and detailed record-keeping are indispensable components of advanced disease management.

- **Regular Hive Inspections:** Frequent and thorough inspections help spot early signs of disease or distress, allowing for timely interventions.
- **Data Analysis:** Using record-keeping to track health trends and outcomes over time, which can help in fine-tuning management practices based on actual apiary data.

Community Engagement and Education: Staying connected with the beekeeping community for the latest research updates and management strategies is vital.

- **Workshops and Seminars:** Regularly participating in educational programs can keep you updated on the latest disease management techniques and research findings.
- **Collaboration:** Working with local beekeeping clubs and extension services can help manage regional disease threats more effectively.

Advanced disease management in beekeeping is a dynamic and complex endeavor that requires ongoing education, vigilance, and adaptation to new challenges and solutions. By embracing these advanced strategies, you not only protect your own hives but also contribute to the health and sustainability of the global bee population. Armed with this knowledge, you're better prepared to tackle the challenges that diseases and pests pose to your precious colonies.

8.3 Making Mead and Other Bee Products

Welcome, enterprising beekeepers! While honey might be the crown jewel of bee products, there's a realm of other treasures that your hive can produce, from the ancient craft of mead making to utilizing beeswax, propolis, and royal jelly. This chapter will explore how to extend the fruits of your labor beyond the hive into delightful and marketable products that celebrate the versatility of beekeeping.

Mead Making: The Brewer's Art

Mead, known as the nectar of the gods, is one of the oldest alcoholic drinks known to man, made from fermenting honey with water and often infused with fruits, spices, herbs, or flowers. Here's how you can embark on the journey of crafting your own mead.

- **Basic Ingredients:** The simplicity of mead lies in its minimal ingredients — honey, water, and yeast. The quality and flavor of the honey significantly impact the taste of the mead, offering an excellent opportunity to highlight unique varietal honeys.
- **The Process:**
    1. **Sanitation:** Begin with thoroughly sanitized equipment to prevent contamination and spoilage.
    2. **Preparation:** Mix honey and water in a large fermenter until the honey is fully dissolved.
    3. **Yeast Addition:** Add yeast once the honey-water mixture, also known as "must," is at room temperature. The type of yeast can affect the flavor and alcohol content of the mead.
    4. **Fermentation:** Allow the mixture to ferment in a cool, dark place. Primary fermentation typically takes about two to six weeks, after which it can be transferred to secondary fermentation to clarify and mature.
    5. **Bottling:** Once fermentation is complete and the mead has cleared, it's time to bottle your mead. Aging can improve the flavor, with some meads aging for years.
- **Creative Variations:** Experiment with adding fruits such as berries, apples, or cherries, or herbs and spices like cinnamon and vanilla during the fermentation process to create unique flavor profiles.

Utilizing Beeswax: A Crafter's Delight

Beeswax, with its delightful scent and versatility, is a valued by-product of beekeeping, used in a variety of crafts and products.

- **Candles:** Beeswax candles burn longer and cleaner than many other types of candles. Making them involves simply melting the wax, adding a wick, and pouring it into molds.

- **Cosmetics:** Beeswax is a common ingredient in natural skincare products, including lip balms, moisturizers, and salves. Its properties help to protect and repair rough, dry skin.
- **Polishes and Waterproofing:** Beeswax is excellent for making natural wood polishes and waterproofing agents for leather.

Propolis: The Bee's Pharmacy

Propolis, used by bees to seal and sterilize their hive, is packed with antibacterial and antifungal properties, making it sought after for various health and cosmetic products.

- **Tinctures:** Propolis tinctures can be made by dissolving propolis in alcohol. This tincture is used for its purported health benefits, including boosting immunity and treating colds and flu.
- **Skin Care Products:** Due to its healing properties, propolis is a valuable addition to ointments and creams, helping to treat wounds, burns, and skin irritations.

Royal Jelly: Liquid Gold

Royal jelly, secreted by nurse bees to feed the queen and young larvae, is renowned for its health benefits and is used in supplements and skincare products.

- **Harvesting:** Collecting royal jelly is labor-intensive, requiring the careful removal of queen cells.
- **Products:** Often sold as dietary supplements, royal jelly is praised for its potential to boost energy and immune system strength. In cosmetics, it's used for its skin-soothing and anti-aging properties.

By branching into these diverse uses of bee products, you can greatly enhance the economic potential of your beekeeping practice while offering natural, high-quality products. Whether you're crafting a batch of golden mead, molding beeswax candles, concocting propolis tinctures, or processing royal jelly, each product not only adds value to your endeavors but also helps in promoting sustainable practices and celebrating the incredible work of bees.

Chapter 9: The Business of Beekeeping

This chapter is designed for those ready to take their passion for beekeeping and turn it into a profitable venture. We'll explore the essential steps to professionalize your operation, effective strategies for marketing and selling your bee products, and the importance of networking within the beekeeping community.

9.1 Turning Beekeeping into a Business:

Learn how to establish a business plan, understand the legalities involved, and manage your finances to ensure your beekeeping business is both sustainable and profitable.

9.2 Marketing and Selling Your Products:

Discover how to effectively market your unique bee products, from honey and wax to propolis and royal jelly, and find the right markets where you can sell them to maximize your profits.

9.3 Networking with Other Beekeepers:

Networking is crucial in the beekeeping business. We'll discuss how connecting with other beekeepers can provide support, enhance your learning, and open up new business opportunities.

This chapter will provide you with the tools and knowledge needed to turn your beekeeping practice into a thriving business, ensuring you can continue your passion for bees while also making a viable living.

9.1 Turning Beekeeping into a Business

If you're ready to transform your beekeeping hobby into a flourishing business, this section is tailored for you. Turning beekeeping into a profitable venture involves more than just passion for bees; it requires strategic planning, understanding market dynamics, and effective business management. Let's navigate the essential steps to professionalize your beekeeping operation and ensure its sustainability and profitability.

Develop a Business Plan: The cornerstone of any successful business is a robust business plan. This plan will serve as a roadmap for your beekeeping business, outlining your objectives, strategies, and financial projections.

- **Market Analysis:** Understand your local market demand for bee products. Are consumers more interested in raw honey, beeswax products, or perhaps medicinal propolis? Identifying what sells best in your community can guide your production focus.
- **Business Goals:** Set clear, achievable goals. Whether it's expanding your apiary, diversifying your product line, or reaching certain revenue targets, having defined goals will keep your business growth on track.

- **Budgeting:** Include a detailed budget that covers startup costs, operational expenses, and forecasts for revenue. Don't forget to factor in the costs of beekeeping supplies, hive maintenance, and any labor you might need.

Legal Considerations: To turn your beekeeping into a legitimate business, you'll need to navigate various legal requirements, which can vary widely depending on your location.

- **Business Registration:** Register your business with the appropriate local authorities. This may include obtaining a business license and setting up the right type of business entity (such as an LLC, sole proprietorship, etc.) to manage liability and tax obligations effectively.
- **Health and Safety Regulations:** Ensure compliance with health and safety regulations related to food production if you plan to sell honey or other consumables. This might involve regular health inspections and specific labeling requirements.
- **Insurance:** Consider obtaining insurance to protect your business from liability, especially if you plan to operate a visitor-friendly apiary or attend public markets.

Production Management: Efficient management of your beekeeping operations is vital for maintaining the health of your bees and the quality of your products.

- **Sustainable Practices:** Adopt sustainable beekeeping practices to ensure the long-term health of your bee colonies. This includes responsible colony management, disease control, and possibly organic certification if it aligns with your market.
- **Quality Control:** Implement strict quality control measures for your products. High-quality, consistent products will help establish your brand's reputation in the market.
- **Scaling Up:** Plan for scaling your operations. This may involve increasing hive numbers, automating certain processes, or hiring additional help during peak seasons.

Marketing and Branding: Developing a strong brand and effective marketing strategies are crucial for capturing and growing your market share.

- **Brand Identity:** Create a compelling brand identity that reflects the quality and uniqueness of your products. This includes a memorable brand name, logo, and cohesive marketing materials.
- **Online Presence:** Establish an online presence with a professional website and active social media accounts to connect with customers and promote your products.
- **Networking:** Build relationships with local businesses and consumers. Attend farmer's markets, join local business associations, or collaborate with local restaurants and shops to get your products to a wider audience.

Customer Engagement: Engaging with your customers can transform one-time buyers into loyal patrons.

- **Educational Outreach:** Educate your customers about the benefits of your products and the importance of sustainable beekeeping. Workshops, open house events at your apiary, or educational content on your website can enhance customer engagement.
- **Feedback Loops:** Establish channels for customer feedback. Listening to customer experiences and suggestions can provide valuable insights for improving your products and services.

Turning beekeeping into a business is an exciting journey that can lead to substantial personal and financial rewards. With careful planning, compliance, and creativity, your beekeeping business can thrive and contribute positively to your community and the environment. So, spread your wings and let your beekeeping business soar!

9.2 Marketing and Selling Your Products

You've nurtured your bees, harvested your products, and now you're ready to step into the marketplace. Marketing and selling your bee products effectively are crucial to turning your beekeeping endeavor from a passion into a profitable business. Let's explore comprehensive strategies for marketing your honey, beeswax, propolis, and other bee-derived delights, ensuring your products find their way into the hearts and homes of consumers.

Understanding Your Market: Before you can sell effectively, you need to understand who your customers are, what they value, and where you can reach them.

- **Market Research:** Conduct thorough market research to identify your target audience. Are they health-conscious consumers, gourmet food lovers, or perhaps local businesses looking for quality ingredients?
- **Competitive Analysis:** Analyze your competition. What are other beekeepers and similar businesses offering? What can you do differently or better? Understanding your unique selling proposition is key.

Branding Your Bee Products: Creating a strong, memorable brand is crucial. Your brand is the promise of quality and reliability that you communicate to your customers.

- **Logo and Design:** Develop a professional logo and attractive packaging that reflects the quality of your products and appeals to your target market. Your design should communicate the natural and sustainable nature of your products.
- **Storytelling:** Share the story of your beekeeping journey through your branding. Customers love to hear about the origin of their products, the sustainability efforts you practice, and the quality of life that your bees enjoy.

Effective Marketing Strategies: With your target market in mind and your branding in place, it's time to deploy effective marketing strategies.

- **Digital Marketing:** Utilize social media platforms to connect with customers, share engaging content about your products, and educate your audience about the benefits of bee products. Consider email marketing to keep your customers informed about new products, promotions, or beekeeping events.
- **Content Marketing:** Develop valuable content such as blogs, videos, and articles that highlight the uses and benefits of your products. SEO (Search Engine Optimization) can help ensure that your online content reaches the right audience.
- **Local Markets and Events:** Participate in local farmers' markets, craft fairs, and food expos. These venues are excellent for making direct sales and building relationships with your community.

Sales Channels: Deciding where and how to sell your products can impact your business's success.

- **Online Sales:** Setting up an online store can reach a broader audience and offer convenience to customers. Platforms like Etsy, Amazon, or your own website are excellent outlets.
- **Retail Partnerships:** Partner with local grocery stores, health food stores, and boutiques to carry your products. Retail partnerships can increase your product visibility and credibility.
- **Direct Sales:** Consider selling directly from your apiary or through local food cooperatives. Direct sales often allow for better profit margins and customer relationships.

Customer Relations and Retention: Maintaining good relationships with your customers ensures repeat business and enhances word-of-mouth marketing.

- **Customer Service:** Provide excellent customer service by being responsive, helpful, and transparent. Handle any complaints swiftly and professionally.
- **Loyalty Programs:** Implement loyalty programs or offer discounts to repeat customers. Such incentives encourage continued business and can attract new customers through referrals.

Legal Considerations: Ensure you adhere to all local, state, and federal regulations regarding the production, labeling, and sale of food products. Proper labeling is critical, especially for allergens and expiry dates, and can help avoid legal issues and build trust with your customers.

Evaluating Performance: Regularly evaluate the effectiveness of your marketing and sales strategies. Which techniques are bringing in the most customers? What feedback are you receiving? Use this data to refine your approach continuously.

By mastering these marketing and sales strategies, you can create a thriving business that not only supports you financially but also promotes the importance of bees in our ecosystem. Your passion for beekeeping can thus be shared with a broader audience, educating and inspiring as it grows.

9.3 Networking with Other Beekeepers

n the world of beekeeping, the adage "no man is an island" couldn't be more true. Networking with fellow beekeepers isn't just about socializing; it's a strategic approach that can significantly enhance both the success of your business and the health of your bees. Engaging with a community of beekeepers allows for the exchange of knowledge, experiences, and resources, fostering a collaborative environment that benefits everyone involved. Let's delve into why networking is crucial and explore effective ways to connect with the beekeeping community.

Understanding the Benefits of Networking: Networking with other beekeepers provides numerous advantages:

- **Shared Knowledge:** Beekeeping practices can vary greatly depending on climate, local flora, and bee species. Sharing experiences and strategies with others can uncover insights you might not find in books or online.
- **Disease and Pest Management:** Learn about the latest methods for managing pests and diseases in your region. Networking can also alert you to new threats and outbreaks, allowing for proactive measures.
- **Resource Sharing:** Sometimes beekeeping requires resources that can be costly or hard to find. Through your network, you might share equipment, buy supplies in bulk together for a discount, or exchange services like hive splitting or queen rearing.
- **Breeding and Genetics:** Networking can open opportunities for accessing genetically diverse bees, enhancing the health and vigor of your colonies.
- **Support and Mentorship:** For new beekeepers, having a mentor is invaluable. Experienced beekeepers can provide guidance, troubleshooting tips, and moral support.

Effective Networking Strategies: Building a network in the beekeeping community involves both in-person and digital strategies.

- **Join Local Beekeeping Clubs and Associations:** Most regions have their own beekeeping clubs that host meetings, workshops, and social events. These are great places to meet fellow beekeepers and learn about local beekeeping challenges and opportunities.

- **Attend Conferences and Workshops:** Larger gatherings like state or national beekeeping conferences are not only educational but also excellent for connecting with a wider network of beekeepers.
- **Participate in Online Forums and Social Media Groups:** Platforms like Beesource, BeeMaster forums, or Facebook groups offer a way to connect with beekeepers worldwide. These forums are great for asking questions, sharing experiences, and staying updated on beekeeping news.
- **Volunteer:** Participating in community outreach programs can help promote beekeeping awareness and education. This can be a rewarding way to meet people interested in beekeeping and to contribute to the community.
- **Collaborations and Partnerships:** Look for opportunities to collaborate on projects or engage in partnerships with universities, agricultural organizations, or environmental groups. These can lead to new business opportunities and ways to contribute to research and conservation efforts.

Maintaining Relationships: Networking is not just about meeting people; it's about cultivating relationships over time.

- **Follow-Up:** After meeting new contacts, follow up with a message or email. Express your appreciation for the conversation and suggest another meeting or check-in.
- **Stay Active:** Regularly participate in meetings and events. Being visible and engaged helps strengthen relationships and keeps you on top of new developments in the community.
- **Offer Help:** Networking is a two-way street. Offering assistance or sharing resources with others can build goodwill and establish you as a valuable member of the community.

Documenting Your Networking Journey: Keep a record of the people you meet and the insights you gain. This can be an invaluable resource as you apply new techniques in your beekeeping practice or seek advice on specific challenges.

Networking in beekeeping is not just beneficial; it's essential. It helps build a supportive community that enhances the resilience and sustainability of your beekeeping endeavors. By engaging with other beekeepers, you enrich your own experience and contribute to the broader goal of promoting healthy bee populations and ecosystems. So step out, connect, and watch as both your bees and your relationships flourish!

Chapter 10: Sustainable Beekeeping Practices

In this chapter we discuss the essentials of sustainable beekeeping—a practice that not only nurtures your bees but also enriches the environment around us. This chapter focuses on implementing eco-friendly techniques that minimize negative impacts on local ecosystems and promote biodiversity. By adopting sustainable practices, beekeepers can play a crucial role in supporting their local environments while enjoying the bountiful rewards of beekeeping. From choosing the right materials and methods to understanding the broader ecological implications of your actions, this chapter provides a comprehensive guide to making beekeeping a force for positive environmental change.

10.1 Eco-Friendly Beekeeping Techniques

As beekeepers, our practices can have profound impacts on the environment. By adopting sustainable methods, we can help ensure the health of our bee colonies and contribute positively to the ecosystem. This comprehensive guide delves into various eco-friendly beekeeping techniques, offering practical advice for those looking to make their beekeeping practices more sustainable.

Understanding Eco-Friendly Beekeeping: Eco-friendly beekeeping involves methods that minimize environmental impact, use sustainable resources, and promote the health and survival of bees within natural ecosystems. This approach is about more than just avoiding harm; it's about actively contributing to a healthier environment.

1. Natural Hive Materials: Choosing the right materials for your hives can significantly impact their sustainability.

- **Sustainable Wood:** Opt for locally sourced, sustainably harvested wood. Look for certifications like FSC (Forest Stewardship Council) to ensure wood is sourced responsibly.
- **Non-Toxic Coatings:** Use natural, non-toxic paints and varnishes to treat wood. Products based on linseed oil or other natural substances not only protect the hive from the elements but also ensure no harmful chemicals leach into the hive or the environment.

2. Chemical-Free Pest Management: Reducing or eliminating the use of synthetic chemicals to control pests and diseases is crucial for maintaining an eco-friendly apiary.

- **Integrated Pest Management (IPM):** IPM focuses on preventative measures, such as maintaining strong, healthy colonies that can naturally resist pests and diseases. Techniques include managing bee space effectively, using physical barriers to prevent mite entry, and encouraging beneficial insects that prey on harmful pests.

- **Organic Treatments:** When treatments are necessary, opt for organic options like oxalic or formic acid, which are less harmful to bees, humans, and the environment compared to synthetic chemicals.

3. Promoting Local Forage and Biodiversity: Enhancing the foraging options available to your bees helps to support not only your colonies but also the surrounding ecosystem.

- **Planting Bee Gardens:** Create a bee-friendly garden with a variety of plants that bloom at different times of the year, providing continuous forage for bees. Focus on native plants, as these are often best suited to both local bees and other pollinators.
- **Supporting Wild Habitats:** Preserve existing natural habitats around your apiary. Protecting areas such as woodlands, wetlands, and meadows supports a wide range of wildlife and plant species, which in turn supports ecological balance.

4. Water Conservation and Management: Bees need water for various functions, but sustainable water management in the apiary can help conserve this vital resource.

- **Water Sources:** Provide bees with water through sustainable means such as rainwater harvesting. Set up shallow water trays with stones or marbles for bees to land on, reducing the risk of drowning.
- **Avoiding Contamination:** Ensure that water sources are clean and free from contaminants. Position water sources away from potential pollutants like agricultural runoff.

5. Energy Efficiency and Renewable Energy: Utilizing renewable energy sources for any powered elements of your apiary can reduce your carbon footprint.

- **Solar Power:** Consider using solar panels to power supplemental heating in hives or lighting in work areas. Solar energy can also be used to power electric fences that protect hives from bears and other predators.
- **Energy-Efficient Tools:** Invest in energy-efficient tools and machinery for honey extraction and processing. This not only reduces energy use but can also decrease operational costs over time.

6. Waste Reduction and Recycling: Minimizing waste in beekeeping operations is both eco-friendly and economically wise.

- **Wax and Propolis Reuse:** Recycle wax from old combs and propolis collected during hive inspections. These can be melted down and used to make new products like candles, balms, and tinctures.
- **Composting:** Organic waste such as plant matter from around your hives can be composted to enrich the soil, promoting healthier plant growth and thus better forage for bees.

7. Education and Advocacy: As beekeepers, we have a unique opportunity to educate others about the importance of sustainable practices and the role bees play in our ecosystem.

- **Community Engagement:** Hold workshops or talks at local schools, libraries, or community centers. Teach people about the importance of bees to our food systems and how they can help support pollinator populations.
- **Policy Advocacy:** Engage with local and national policymakers to advocate for bee-friendly policies. This could include measures to protect habitat, limit pesticide use, or support organic farming practices.

By integrating these eco-friendly beekeeping techniques into your practice, you not only enhance the health and productivity of your own hives but also contribute to broader environmental sustainability. This holistic approach not only benefits the biodiversity and resilience of local ecosystems. Such a comprehensive approach to beekeeping ensures that you are not just a keeper of bees but also a guardian of the environment, fostering a sustainable future for both bees and humans alike. Through mindful practices and continuous learning, we can make a significant positive impact—one hive at a time.

10.2 Minimizing Impact on Local Ecosystems

hen managing hives, the responsibility extends beyond the bees to the local ecosystems where they thrive. Beekeeping can have significant impacts, both positive and negative, on local flora and fauna. This chapter explores how to minimize the negative impacts of beekeeping on local ecosystems, ensuring that your practices promote environmental harmony and contribute positively to the surrounding biodiversity.

Understanding Ecosystem Impact: Beekeeping interacts with local ecosystems in various ways, from competition with native pollinators to the spread of diseases and the potential for resource depletion. Understanding these interactions is crucial for developing practices that reduce negative impacts.

- **Competition with Native Pollinators:** Honey bees can compete with native pollinators for limited resources. This competition can potentially stress local ecosystems, especially in areas where resources are scarce or native pollinator populations are vulnerable.
- **Potential Disease Transmission:** Bees can be vectors for diseases that may spread to wild bee populations, posing significant risks to native species.
- **Resource Depletion:** Overstocking an area with hives can lead to the depletion of nectar and pollen, leaving insufficient food for both managed honey bees and native pollinators.

Strategies for Minimizing Ecosystem Impact:
1. **Responsible Hive Placement:**

- **Assess Carrying Capacity:** Before placing hives, assess the floral resources available and the existing pollinator populations to avoid overstocking. Use local flora surveys and consult with ecological experts to gauge the area's carrying capacity.
- **Maintain Distance:** Keep hives away from sensitive habitats where protected or endangered species of pollinators are known to exist, to reduce competition and disease transmission risks.

2. **Promote Floral Diversity:**
   - **Plant Native Species:** Enhance the foraging opportunities for all pollinators by planting native flowers, shrubs, and trees that bloom at different times throughout the year. This not only supports a diverse range of pollinators but also helps ensure that your bees have access to a consistent source of food.
   - **Avoid Monocultures:** Planting a variety of plants can prevent the monoculture effect, which can be harmful if bees are only exposed to one type of pollen or nectar source.

3. **Disease Management and Biosecurity:**
   - **Regular Health Checks:** Monitor your bees regularly for signs of disease and manage them promptly to prevent any spread to wild populations.
   - **Biosecurity Measures:** Implement strict biosecurity measures, such as sanitizing equipment and controlling apiary access, to prevent the introduction and spread of pathogens.

4. **Use of Native Bees:**
   - **Support Native Bee Populations:** Where appropriate, consider cultivating or supporting habitats for native bees alongside honey bees. Native bees can be more effective pollinators for certain crops and flowers, and supporting their populations helps maintain ecological balance.

5. **Educate and Collaborate:**
   - **Community Education:** Take an active role in educating the community about the importance of biodiversity and the role of different pollinators within the ecosystem.
   - **Collaborate with Conservationists:** Work with local conservation groups, wildlife experts, and botanical gardens to align your beekeeping practices with local conservation efforts.

6. **Sustainable Resource Management:**
   - **Water Conservation:** Implement water conservation practices in your apiary to minimize the impact on local water resources. Use water-efficient methods for any apiary-related activities.

- **Minimal Intervention:** Adopt a minimal intervention approach in your beekeeping practices. This includes reducing the use of chemicals, letting bees build natural comb structures, and using organic methods whenever possible.

7. **Monitoring and Adapting:**
    - **Impact Assessment:** Regularly assess the impact your apiary has on the local ecosystem. Look for signs of stress in local flora and fauna, and adjust your practices accordingly.
    - **Feedback Mechanisms:** Establish mechanisms for feedback from the local community and environmental groups to continually improve your practices and reduce impacts.

Minimizing the impact of beekeeping on local ecosystems is not only about safeguarding the environment but also about ensuring the sustainability of beekeeping itself. By adopting these strategies, you help create a balance where your beekeeping practices contribute positively to local biodiversity, supporting the health of your bees and the vibrancy of the ecosystems they inhabit. Engaging in sustainable beekeeping practices ensures that your impact extends beyond the hive, fostering a healthier planet for future generations.

10.3 Promoting Biodiversity Through Beekeeping

In this vital section of our beekeeping journey, we delve into how beekeeping practices can actively promote biodiversity. Biodiversity encompasses the variety of all biological species, playing a crucial role in ecosystem health and resilience. By adopting beekeeping practices that enhance biodiversity, we not only support our bee populations but also contribute to the wider ecological community, enhancing the health and sustainability of local and global environments.

Why Biodiversity Matters in Beekeeping:

Biodiversity ensures ecological stability and productivity. Healthy ecosystems with diverse species offer resilience against environmental stresses and changes, such as climate fluctuations and disease outbreaks. For beekeepers, promoting biodiversity means fostering a range of plant species that offer bees varied and continuous nutrition, and supporting other pollinators and wildlife that contribute to a balanced ecosystem.

Strategies for Promoting Biodiversity:

1. **Diverse Planting Practices:**
    - **Season-Long Forage:** Cultivate a mix of plants that flower at different times throughout the year to provide bees with a steady supply of food.
    - **Native Plant Focus:** Use native plants in apiary surroundings. These plants are better adapted to the local environment and provide optimal support for native bee species and other local wildlife.

- **Avoid Invasive Species:** Be mindful of plant selection to avoid species known to be invasive or harmful to the local ecosystem.

2. **Habitat Conservation:**
    - **Natural Habitat Areas:** Maintain or create areas that provide natural habitats for various wildlife, which can also serve as additional foraging sites for bees.
    - **Wildflower Meadows:** Establish wildflower meadows which are not only excellent for bees but also support a wide variety of pollinators and insects.

3. **Sustainable Hive Management:**
    - **Chemical-Free Practices:** Limit or eliminate the use of chemicals in managing hives. Opt for natural pest and disease management strategies to reduce environmental toxins.
    - **Eco-Friendly Materials:** Choose sustainable materials for hive construction and maintenance to minimize environmental impact.

4. **Water Resource Management:**
    - **Water Conservation:** Implement systems to collect and use rainwater for apiary operations to reduce water consumption.
    - **Safe Water Features:** Provide clean, safe watering points for bees and other wildlife, ensuring they are designed to prevent drowning and are easy to maintain.

5. **Community Engagement and Education:**
    - **Educational Programs:** Lead or participate in educational initiatives that teach the community about the importance of biodiversity and sustainable beekeeping.
    - **Collaborations:** Work with local environmental groups, schools, and community organizations to promote and implement biodiversity-friendly practices.

6. **Research and Monitoring:**
    - **Citizen Science:** Engage in or support research efforts that monitor the effects of beekeeping on local biodiversity. Citizen science projects can provide valuable data on how different practices affect the environment.
    - **Ongoing Evaluation:** Regularly assess the impact of your beekeeping practices on local biodiversity and adjust practices based on findings to enhance ecological benefits.

Benefits of Biodiversity in Beekeeping:
- **Enhanced Bee Health and Productivity:** A diverse environment provides bees with a robust and nutritious diet, improving colony health and increasing honey production.
- **Resilience to Pests and Diseases:** Biodiverse ecosystems are less likely to experience the outbreaks of pests and diseases common in monocultural or less diverse environments.

- **Support for Wider Ecological Health:** By supporting biodiversity, beekeepers help sustain a variety of life forms, contributing to the overall health of the ecosystem, which in turn supports sustainable agriculture and healthy communities.

Promoting biodiversity through beekeeping is not merely about enhancing the number of species in a given area but about enhancing the quality of life for all species, including our own. By adopting practices that support diverse plant life and wildlife, beekeepers can play a pivotal role in sustaining and enriching the environments in which they operate. This holistic approach not only benefits the bees and their keepers but also the planet as a whole, fostering a healthier, more sustainable world for future generations.

Chapter 11: The Future of Beekeeping

As we look to the horizon, beekeeping faces a dynamic landscape filled with rapid innovations, evolving challenges, and the critical need for adaptation due to climate change. This chapter explores the forefront of beekeeping technology, identifies emerging opportunities alongside the challenges, and discusses strategies for adapting practices in response to global climate shifts. We'll delve into how technological advancements are revolutionizing the way we manage hives, the obstacles that threaten to disrupt bee populations, and the adaptive measures necessary to ensure the sustainability and resilience of beekeeping in the future. Through understanding and action, beekeepers can navigate these changes, benefiting both their colonies and the broader ecosystem.

11.1 Innovations in Beekeeping Technology

In the evolving world of beekeeping, technology plays an increasingly pivotal role, pushing the boundaries of what's possible in hive management, disease control, and productivity enhancements. This section explores the cutting-edge innovations transforming the beekeeping landscape, offering beekeepers new tools and techniques to improve their craft and increase the sustainability of their operations.

Remote Hive Monitoring Systems:

One of the most significant advancements in beekeeping technology is the development of remote hive monitoring systems. These systems use sensors placed within or near hives to collect data on a variety of parameters including temperature, humidity, hive weight, and even sound frequencies emitted by the bees.

- **Temperature and Humidity Sensors:** These help beekeepers monitor the internal conditions of the hive, ensuring they remain optimal for bee health and activity. Alerts can be set up to notify the beekeeper when conditions deviate from ideal ranges, allowing for quick adjustments.
- **Hive Weight Monitoring:** By tracking changes in hive weight, beekeepers can estimate honey production and monitor the health of the colony. Significant weight loss might indicate issues such as a decrease in foraging activity or health problems within the hive.
- **Acoustic Monitoring:** Sophisticated algorithms analyze the sounds that bees produce. Changes in buzzing patterns can indicate various states of colony health, from queen presence to the stress levels within the hive.

Automated Hive Management Tools:

Automation technology is beginning to permeate the beekeeping industry, offering tools that reduce the labor intensity of certain tasks.

- **Automated Honey Extractors:** These devices streamline the process of extracting honey by automating the uncapping, spinning, and filtering stages, thereby increasing efficiency and reducing the beekeeper's workload.
- **Robotic Frame Handlers:** Emerging technologies include robotic systems designed to handle frames during inspections or harvesting, minimizing the disruption to bees and reducing the labor required for hive management.

Advanced Disease Detection and Management:

Technology is also revolutionizing how bee diseases and pests are detected and managed, offering more precise and early detection methods.

- **Diagnostic Kits:** Portable diagnostic kits that use molecular techniques like PCR (Polymerase Chain Reaction) can detect pathogens in bee colonies on-site, providing results much faster than traditional lab-based tests. This rapid diagnosis allows for quicker responses to health threats.
- **Integrated Pest Management (IPM) Apps:** Mobile applications that integrate data from various sources to provide beekeepers with real-time advice on pest management, tailored to their specific environmental conditions and hive statuses.

AI and Machine Learning:

Artificial intelligence (AI) and machine learning are starting to be applied in beekeeping for various purposes, from improving breeding programs to predicting colony collapses.

- **Predictive Analytics:** AI algorithms can analyze data collected from hive sensors to predict potential problems before they become serious, such as predicting the likelihood of a disease outbreak or a potential swarming event.
- **Genetic Analysis Tools:** AI-driven tools analyze genetic data from bees to help in breeding programs, selecting traits that enhance disease resistance or productivity.

Drone Technology:

Drones, although more commonly associated with other forms of agriculture, are finding their place in beekeeping, particularly in large-scale operations.

- **Crop Pollination:** Drones are being experimented with to assist in pollinating crops, especially in areas where bee populations are declining.
- **Habitat Assessment:** Drones equipped with cameras can be used to assess the health of the flora around bee habitats, providing beekeepers with detailed information about the availability and health of bee forage areas.

Blockchain for Traceability:

Blockchain technology is being used to enhance the traceability of honey and other bee products, reassuring consumers about the quality and origin of the products they purchase.

- **Supply Chain Transparency:** Blockchain can securely track each step of the honey production and distribution process, from hive to table, ensuring that the product is pure, unadulterated, and sustainably sourced.

These technological innovations are reshaping the field of beekeeping, making it more efficient, sustainable, and productive. By integrating these technologies into their practices, beekeepers can not only enhance their operational efficiency but also contribute to the broader goal of maintaining healthy bee populations and ecosystems. As we continue to face global challenges like climate change and biodiversity loss, these tools will become increasingly crucial in ensuring the resilience and sustainability of beekeeping practices worldwide.

11.2 Challenges and Opportunities Ahead

In the ever-evolving world of beekeeping, the road ahead is marked by both significant challenges and promising opportunities. As beekeepers, understanding these dynamics is crucial for adapting practices and strategies to ensure the sustainability and growth of our operations. This section explores some of the major challenges and opportunities that lie ahead for beekeepers around the world.

Challenges Facing Beekeepers:

1. **Climate Change:**
   - **Impact on Foraging Patterns:** Climate change is altering the blooming times of many plants, which can disrupt the availability of food for bees. These changes may force bees to adapt quickly to new conditions that may not always be conducive to their health or productivity.
   - **Weather Extremities:** Increasing occurrences of extreme weather conditions—such as droughts, heavy rains, and extreme temperatures—pose direct threats to bee colonies, impacting their survival and the viability of beekeeping operations.
2. **Pesticide Use:**
   - **Chemical Exposure:** Despite growing awareness, the use of pesticides continues to be a significant threat to bees, affecting their health and the overall environmental balance. Pesticide exposure can weaken bee immune systems, making them more susceptible to diseases and pests.

- **Regulatory Challenges:** Navigating the regulatory landscapes regarding pesticide use in agriculture remains a challenge, as many harmful chemicals are still in use and affect bee populations.

3. **Disease and Pest Management:**
    - **New and Emerging Threats:** The spread of diseases and pests continues to be a major concern, with issues like Varroa mites and hive beetles still prevalent. Emerging threats are continuously monitored, requiring beekeepers to stay informed and prepared to implement new management strategies.
    - **Resistance to Treatments:** Over time, pests and pathogens can develop resistance to traditional treatments, making them less effective and requiring the development of new solutions.

Opportunities for Beekeepers:

1. **Technological Advancements:**
    - **Innovative Monitoring Tools:** Technologies such as hive monitoring systems offer beekeepers unprecedented real-time data on hive health, allowing for more precise and effective management.
    - **Breeding Programs:** Advances in genetic research can help develop more resilient bee strains that are better equipped to handle environmental stresses and resist diseases.

2. **Increased Public Awareness:**
    - **Support for Bee-friendly Practices:** There is a growing public and governmental support for bee-friendly practices, including urban beekeeping and the planting of bee-friendly flora in public spaces.
    - **Educational Opportunities:** The increased interest in beekeeping as a hobby provides a significant opportunity for experienced beekeepers to offer workshops, courses, and consultations, potentially opening new revenue streams.

3. **Sustainability Initiatives:**
    - **Eco-friendly Certification:** Opportunities for obtaining certifications for organic and sustainable beekeeping practices can allow beekeepers to access niche markets and premium pricing for their products.
    - **Collaborations with Environmental Groups:** Partnering with conservation organizations can help beekeepers gain access to resources, funding, and platforms to promote sustainable practices.

4. **Market Expansion:**

- **Diversification of Products:** Beyond honey, there is increasing demand for other bee-related products such as beeswax, propolis, and royal jelly. Developing these product lines can help beekeepers tap into new markets.
- **Global Markets:** Expanding sales to global markets can be facilitated by the internet and improvements in shipping and logistics, offering beekeepers the chance to promote their products on a larger scale.

The future of beekeeping requires navigating these challenges with innovative solutions and taking full advantage of emerging opportunities. By staying informed, adaptable, and proactive, beekeepers can not only overcome these hurdles but also thrive, ensuring the health of their colonies and contributing positively to global biodiversity and environmental sustainability.

11.3 Adapting to Climate Change

As global climates continue to shift, beekeepers face a myriad of challenges that require innovative and forward-thinking solutions. Climate change is not just altering the landscape; it's reshaping the very environment in which bees thrive. From changing flowering patterns to extreme weather events, beekeepers must navigate these turbulent waters to ensure the sustainability and health of their colonies. This chapter explores how beekeepers can adapt their practices to the realities of climate change, ensuring their operations are resilient and can flourish in the face of environmental shifts.

Understanding the Impact of Climate Change on Beekeeping:

Climate change affects beekeeping in several profound ways:

- **Altered Flowering Times:** As temperatures warm, the phenology of plants shifts, leading to earlier or later blooms. This can disrupt the synchronization between when flowers produce pollen and nectar and when bees are active and need these resources.
- **Extreme Weather Conditions:** Increased frequency and intensity of extreme weather events—such as storms, droughts, and heatwaves—can directly threaten bee colonies. Such conditions can lead to habitat loss, reduced foraging opportunities, and increased stress on bees, which in turn affects their health and productivity.
- **Habitat Loss:** Gradual changes in climate can lead to the alteration of habitats that bees and other pollinators rely on. This can reduce the availability of native flora that is essential for bee nutrition and health.

Strategies for Adapting to Climate Change:

1. **Enhanced Monitoring and Management:**
    - **Data-Driven Decisions:** Utilize climate data and predictive models to anticipate changes in weather patterns and flowering times. This knowledge can help in making informed decisions about colony management and migratory beekeeping practices.
    - **Robust Monitoring Systems:** Implement advanced monitoring systems that can provide real-time data on hive conditions, helping to swiftly address any issues that arise due to sudden climatic changes.
2. **Diversifying Forage Sources:**
    - **Cultivating Resilient Plants:** Plant a variety of nectar and pollen sources that bloom at different times throughout the year, ensuring continuous availability regardless of shifts in individual plant flowering times.
    - **Creating Microclimates:** Develop microclimates within your apiary, such as windbreaks, shaded areas, and moisture-retaining landscapes, to mitigate the effects of extreme temperatures and weather conditions.
3. **Colony Stress Reduction:**
    - **Water Management:** Ensure adequate and consistent water sources for bees, particularly during hot and dry periods. Techniques like providing shaded water stations or automatic waterers can prevent dehydration and overheating.
    - **Hive Insulation:** Adapt hive designs to improve insulation against extreme heat and cold, potentially incorporating materials that can help regulate the internal temperature of hives.
4. **Community and Legislative Action:**
    - **Advocacy and Policy:** Engage with agricultural and environmental groups to advocate for policies that address the broader impacts of climate change on agriculture and beekeeping.
    - **Community Education:** Educate local communities and fellow beekeepers about the impacts of climate change on pollinators and the importance of collective action to mitigate these effects.
5. **Breeding and Genetic Diversity:**
    - **Resilient Bee Strains:** Invest in breeding programs focused on enhancing the resilience of bees to environmental stresses like heat and humidity.
    - **Genetic Diversity:** Maintain high genetic diversity within your colonies to increase their overall resilience to diseases, pests, and environmental stresses.

Building Resilient Systems:
- **Long-term Planning:** Develop long-term adaptation strategies that anticipate further climatic shifts, focusing on sustainability and resilience in beekeeping practices.
- **Innovative Research:** Collaborate with research institutions to study the impacts of climate change on bee health and productivity and to develop new technologies and methods for adaptation.

Adapting to climate change is not just a challenge; it's an imperative for the continued success and sustainability of beekeeping. By understanding the specific ways in which climate change can affect beekeeping and by implementing strategic adaptations, beekeepers can protect their colonies and even thrive in a changing world. This proactive approach will not only safeguard the livelihood of beekeepers but also ensure the preservation of vital pollinator populations, securing ecological health and agricultural productivity for future generations.

Chapter 12: Collect Your Bonus

Unlock Year-Round Success with Your Beehive! This bonus chapter introduces the *Beekeeper Calendar*, a comprehensive guide tailored for each month of the year. For every month, discover the specific challenges you might face, the key activities you should perform, and practical tips to enhance your beekeeping practice. Whether it's managing swarms in May or preparing for winter in November, this calendar is your roadmap to a thriving hive. **Access this invaluable tool by scanning the QR code provided, and start planning your beekeeping activities with confidence!**

Ensure the Health of Your Hive at Every Inspection! Dive into the second bonus, the Bee Health Checklist for Inspection. This checklist is a crucial tool for maintaining the health and productivity of your hives. It provides a detailed list of indicators to watch for during each inspection, helping you identify potential issues early and take appropriate action. From checking for signs of disease to monitoring brood patterns, this checklist covers it all. Download this essential resource by scanning the accompanying QR code, and keep your bees buzzing with health year-round!

We'd Love to Hear From You!

"Enjoyed your journey into beekeeping? Please scan the QR code to leave a review and share your experience!"!

Conclusions

As we reach the conclusion of our journey through "Beekeeping for Beginners," it's clear that beekeeping is more than just a hobby or a profession; it is a pivotal element in the broader dialogue about sustainability, biodiversity, and environmental stewardship. Throughout this book, we have explored various facets of beekeeping, from the basics of setting up your first hive to the complexities of managing bee health and adapting to the challenges posed by a changing climate. This concluding chapter aims to weave these threads together, highlighting the crucial role beekeepers play in our ecosystems and providing a forward-looking perspective on how this role may evolve in the future.

Reflecting on What We've Learned

We began with the basics—understanding why we keep bees and how they benefit the environment and us. This foundational knowledge is crucial for anyone entering the field of beekeeping, providing the motivation to pursue this endeavor with passion and respect for the creatures we manage. As we progressed into the practical aspects of beekeeping, such as selecting equipment, choosing a site, and installing and managing hives, the focus was always on sustainable and ethical practices.

The discussion on the life cycle of bees, their behavior, and social structure, along with detailed insights into various hive types and bee species, underscored the complexity and the wonders of bee biology. Understanding these elements is vital for effective colony management and ensures that beekeepers can make informed decisions that support the health and productivity of their bees.

Challenges and Opportunities Ahead

The journey through this book also acknowledged the challenges beekeepers face today—climate change, disease management, and the need for technological adaptation were recurrent themes. These challenges are significant, yet they come with opportunities to innovate and improve the ways we interact with our environment. The sections on advanced beekeeping topics and the business of beekeeping opened up avenues for beekeepers to expand their operations and impact, illustrating that beekeeping is not static but a dynamic field ripe with potential for growth and development.

The Role of Beekeepers in Promoting Biodiversity

One of the most critical roles that beekeepers play is that of an environmental steward. As discussed in the chapters on promoting biodiversity and sustainable practices, beekeepers have a unique opportunity to positively impact their local ecosystems. By adopting practices that enhance floral diversity, reduce pesticide use, and support native wildlife, beekeepers can help sustain the ecological balances necessary for both agricultural and natural landscapes to thrive.

Adapting to the Future

Looking forward, the adaptability of beekeepers will be tested as environmental conditions continue to evolve. The strategies outlined for adapting to climate change are just the beginning. Ongoing education, community engagement, and advocacy for supportive policies will be essential for the beekeeping community to navigate the uncertainties of the future effectively.

Continuing Education and Community Engagement

As we conclude, it is important to emphasize the value of lifelong learning and community involvement. Beekeeping, like any agricultural practice, benefits greatly from an engaged and informed community. Participation in beekeeping clubs, online forums, and continuous learning opportunities can help beekeepers stay informed about new research, innovative practices, and regulatory changes that impact beekeeping.

This book has aimed to equip you with the knowledge and tools needed to start and sustain a successful beekeeping operation. The journey does not end here, however. Each colony you manage is an opportunity to learn and grow. Each challenge faced is a chance to innovate and adapt. Beekeeping is a profound responsibility—a commitment to care for one of nature's most vital pollinators. As you continue on your beekeeping journey, remember that your practices can contribute to a larger global effort to preserve and enhance our environmental heritage for future generations.

In closing, whether you are a novice beekeeper starting your first hive or an experienced apiarist looking to expand your knowledge and impact, remember that your efforts are vital. You are not only producing honey or other bee products; you are actively participating in the stewardship of our planet.

References

Images

1. Queen
    - Source: Wikimedia Commons
    - Creator: HoraMora
    - License: CC BY 4.0 (Creative Commons Attribution 4.0 International)
    - License Details: https://creativecommons.org/licenses/by-sa/2.5/deed.en
    - Accessed: May 10, 2024
    - Image Link: https://commons.wikimedia.org/wiki/File:Reina_amarilla.JPG

2. Brown and Black Bees
    - Source: Pickwizard
    - Creator: HoraMora
    - License: CC0 (Creative Commons Zero)
    - License Details: https://pikwizard.com/cc0-license/
    - Accessed: May 10, 2024
    - Image Link: https://pikwizard.com/photo/brown-and-black-bees/ffe6d8eb3d56a55f0c4eca3386640dc6/

3. Drone
    - Source: Wikimedia Commons
    - Creator: Waugsberg
    - License: Attribution-Share Alike 3.0 Unported license.
    - License Details: https://creativecommons.org/licenses/by/3.0/deed.en
    - Accessed: May 10, 2024
    - Image Link: https://commons.wikimedia.org/wiki/File:Drone_25a.jpg

4. Apis mellifera
    - Source: Wikimedia Commons
    - Creator: Ivar Leidus
    - License: Creative Commons Attribution-Share Alike 4.0 International
    - License Details: https://creativecommons.org/licenses/by-sa/4.0/deed.en
    - Accessed: May 10, 2024
    - Image Link: https://commons.wikimedia.org/wiki/File:Apis_mellifera_-_Brassica_napus_-_Valingu.jpg

5. Apis Cerana:
    - Source: Wikimedia Commons
    - Creator: Peterwchen
    - License: Creative Commons Attribution-Share Alike 4.0 International
    - License Details: https://creativecommons.org/licenses/by-sa/4.0/deed.en
    - Accessed: May 10, 2024
    - Image Link: https://commons.wikimedia.org/wiki/File:Apis_cerana.jpg

6. Apis Dorsata:
    - Source: Wikimedia Commons
    - Creator: Peterwchen
    - License: Creative Commons Attribution-Share Alike 4.0 International
    - License Details: https://creativecommons.org/licenses/by-sa/4.0/deed.en
    - Accessed: May 10, 2024
    - Image Link: https://commons.wikimedia.org/wiki/File:Bidens-Apis_dorsata-pollen_baskets.jpg

7. Apis Florea:
    - Source: Wikimedia Commons
    - Creator: Uajith
    - License: Creative Commons Attribution-Share Alike 4.0 International
    - License Details: https://creativecommons.org/licenses/by-sa/4.0/deed.en
    - Accessed: May 10, 2024
    - Image Link: https://commons.wikimedia.org/wiki/File:Apis_florea_Bangalore.jpg

8. Beekeeping:
    - Source: Wikimedia Commons
    - Creator: Natiolnal Agricultural Technology Institute
    - License: Creative Commons Attribution-Share Alike 4.0 International
    - License Details: https://creativecommons.org/licenses/by-sa/4.0/deed.en
    - Accessed: May 10, 2024
    - Image Link: https://commons.wikimedia.org/wiki/File:INTA_-_productoras_ap%C3%ADcolas_(1).jpg

9. Smoker:
    - Source: Wikimedia Commons
    - Creator: Peter Grima
    - License: Creative Commons Attribution-Share Alike 2.0 International
    - License Details: https://creativecommons.org/licenses/by-sa/2.0/deed.en
    - Accessed: May 10, 2024
    - Image Link: https://commons.wikimedia.org/wiki/File:Smoke_gun.jpg

10. Cleaning the top of the Frame:
    - Source: Wikimedia Commons
    - Creator: Joe DeLuca
    - License: Creative Commons Attribution-Share Alike 2.0 International
    - License Details: https://creativecommons.org/licenses/by-sa/2.0/deed.en
    - Accessed: May 10, 2024
    - Image Link: https://commons.wikimedia.org/wiki/File:Cleaning_the_Top_of_the_Frames.jpg

11. Honey Extraction:
    - Source: Wikimedia Commons
    - Creator: Dunkle Biene
    - License: Creative Commons Attribution-Share Alike 2.0 International
    - License Details: https://creativecommons.org/licenses/by-sa/2.0/deed.en
    - Accessed: May 10, 2024
    - Image Link: https://commons.wikimedia.org/wiki/File:Honey_extracting.jpg

12. Bee Feeder:
    - Source: Wikimedia Commons
    - Creator: Simon Spped
    - License: Creative Commons CC0 1.0 Universal Public Domain Dedication
    - License Details: https://creativecommons.org/publicdomain/zero/1.0/deed.en
    - Accessed: May 10, 2024
    - Image Link: https://commons.wikimedia.org/wiki/File:BeeFeederLHist.JPG

13. Beekeeping bee brush:
    - Source: Wikimedia Commons
    - Creator: Robert Engelhardt
    - License: Creative Commons Attribution-Share Alike 3.0 Unported license.
    - License Details: https://creativecommons.org/licenses/by-sa/3.0/deed.en
    - Accessed: May 10, 2024
    - Image Link: https://commons.wikimedia.org/wiki/File:Beekeeping_bee_brush.jpg

14. Langstroth Hive:
    - Source: Wikimedia Commons
    - Creator: Koffr
    - License: Creative Commons Attribution-Share Alike 2.5 Generic license.
    - License Details: https://creativecommons.org/licenses/by-sa/2.5/deed.en
    - Accessed: May 10, 2024
    - Image Link: https://commons.wikimedia.org/wiki/File:Langstroth-nastavek.jpg

15. Langstroth Frame:
    - Source: Wikimedia Commons
    - Creator: Luc Viatour
    - License: Creative Commons Attribution-Share Alike 3.0 Unported license..
    - License Details: https://creativecommons.org/licenses/by-sa/3.0/deed.en
    - Accessed: May 10, 2024
    - Image Link: https://commons.wikimedia.org/wiki/File:Langstroth_Frames.jpg

16. Tob Bar Hive:
    - Source: Wikimedia Commons
    - Creator: Maja Dumat
    - License: Creative Commons Attribution 2.0 Generic license.
    - License Details: https://creativecommons.org/licenses/by/2.0/deed.en
    - Accessed: May 10, 2024
    - Image Link: https://commons.wikimedia.org/wiki/File:Top-Bar-Hive_(34067377093).jpg

17. Tob Bar Hive:
    - Source: Wikimedia Commons
    - Creator: Maja Dumat
    - License: Creative Commons Attribution 2.0 Generic license.
    - License Details: https://creativecommons.org/licenses/by/2.0/deed.en
    - Accessed: May 10, 2024
    - Image Link: https://commons.wikimedia.org/wiki/File:Blick_in_die_Top-Bar-Hive_(34746297681).jpg

18. Warre Hive:
    - Source: Wikimedia Commons
    - Creator: Bigbearomaha
    - License: Creative Commons Attribution 4.0 International license.
    - License Details: https://creativecommons.org/licenses/by-sa/4.0/deed.en
    - Accessed: May 10, 2024
    - Image Link: https://commons.wikimedia.org/wiki/File:Modwarre1.jpg

19. Warre Hive:
    - Source: Wikimedia Commons
    - Creator: Wyatt Tyrone Smith
    - License: https://creativecommons.org/licenses/by-sa/4.0/deed.en
    - License Details: https://creativecommons.org/licenses/by-sa/4.0/deed.en
    - Accessed: May 10, 2024
    - Image Link: https://commons.wikimedia.org/wiki/File:Flow_Hive_bee_hive.png

20. Hive in the field:
    - Source: Wikimedia Commons
    - Creator: Lewis Clarke
    - License: Creative Commons Attribution-Share Alike 2.0 Generic license.

- License Details: https://creativecommons.org/licenses/by-sa/2.0/deed.en
- Accessed: May 10, 2024
- Image Link: https://commons.wikimedia.org/wiki/File:London_,_Sutton_-_Mayfield_Lavender_Fields,_Bee_Hive_-_geograph.org.uk_-_4069024.jpg

21. Bee Swarming:
    - Source: Wikimedia Commons
    - Creator: Mark Osgatharp
    - License: Creative Commons Attribution-Share Alike 3.0 Unported license.
    - License Details: https://creativecommons.org/licenses/by-sa/3.0/deed.en
    - Accessed: May 10, 2024
    - Image Link: https://commons.wikimedia.org/wiki/File:HoneyBeeSwarm.JPG

22. Bee Swarming:
    - Source: Wikimedia Commons
    - Creator: Pacific Southwest region
    - License: Creative Commons Attribution 2.0 Generic license.
    - License Details: https://creativecommons.org/licenses/by/2.0/deed.en
    - Accessed: May 10, 2024
    - Image Link: https://commons.wikimedia.org/wiki/File:Honey_Bees_Swarm_(8159050858).jpg

23. Beekeeper with Brush :
    - Source: Wikimedia Commons
    - Creator: US Department of Agriculture
    - License: The license below from the metadata of the image (tag "IFD0:Copyright" contained "Public Domain"). The license visible at Flickr was "Public Domain Mark".
    - Accessed: May 10, 2024
    - Image Link: https://commons.wikimedia.org/wiki/File:20140905-AMS-LSC-0334_(14965637958).jpg

24. Beekeeping using Smoker:
    - Source: Wikimedia Commons
    - Creator: Ich
    - License: Creative Commons Attribution-Share Alike 4.0 International license
    - License Details: https://creativecommons.org/licenses/by-sa/4.0/deed.en
    - Accessed: May 10, 2024
    - Image Link: https://commons.wikimedia.org/wiki/File:Beekeeper_using_bee_smoker.jpg

25. Honey Extractor:
    - Source: Wikimedia Commons
    - Creator: Maja Dumat
    - License: Creative Commons Attribution 2.0 Generic license
    - License Details: https://creativecommons.org/licenses/by/2.0/deed.en
    - Accessed: May 10, 2024
    - Image Link: https://commons.wikimedia.org/wiki/File:4-Waben-Schleuder.jpg

26. Honey Extractor:
    - Source: Wikimedia Commons
    - Creator: Roberto Verzo
    - License: Creative Commons Attribution 2.0 Generic license
    - License Details: https://creativecommons.org/licenses/by/2.0/deed.en
    - Accessed: May 10, 2024
    - Image Link: https://commons.wikimedia.org/wiki/File:Honigschleuder.jpg

27. Honey Filtering:
    - Source: Wikimedia Commons
    - Creator: US Department of Agriculture
    - License: the license below from the metadata of the image (tag "IFD0:Copyright" contained "Public Domain"). The license visible at Flickr was "Public Domain Mark"
    - Accessed: May 10, 2024
    - Image Link: https://commons.wikimedia.org/wiki/File:20140910-ARS-LSC-01352_(15270094525).jpg

28. Varroa Mites In Drone Larve:
    - Source: Wikimedia Commons
    - Creator: Forrest and Kim Starr
    - License: Creative Commons Attribution License
    - License Details: http://www.starrenvironmental.com/imageusepolicy/
    - Accessed: May 10, 2024
    - Image Link: https://commons.wikimedia.org/wiki/File:Starr-180406-0696-Prosopis_pallida-bee_monitoring_training_varroa_mites_in_drone_larvae-HDOA_Hilo-Hawaii_(27498162628).jpg

29. Varroa Mites In Drone Larve:
    - Source: Wikimedia Commons
    - Creator: Denise Anderson
    - License: Creative Commons Attribution 3.0 Unported license.
    - License Details: https://creativecommons.org/licenses/by/3.0/deed.en
    - Accessed: May 10, 2024
    - Image Link: https://commons.wikimedia.org/wiki/File:CSIRO_ScienceImage_7306_A_European_honey_bee_prepupa_with_varroa_mites.jpg

30. Small Beetles:
    - Source: Wikimedia Commons
    - Creator: Denise Anderson
    - License: Creative Commons Attribution 3.0 Unported license.
    - License Details: https://creativecommons.org/licenses/by/3.0/deed.en
    - Accessed: May 10, 2024
    - Image Link: https://commons.wikimedia.org/wiki/File:CSIRO_ScienceImage_1888_Small_Beetles_in_a_Hive.jpg

31. Paenibacillus larvae
    - Source: Wikimedia Commons
    - Creator: Tanarus
    - License: Creative Commons Attribution 3.0 Unported license.
    - License Details: https://creativecommons.org/licenses/by/3.0/deed.en
    - Image Link: https://commons.wikimedia.org/wiki/File:Paenibacillus_larvae.jpg
    - Accessed: May 10, 2024

32. European foulbrood
    - Source: Wikimedia Commons
    - Creator: Výzkumný ústav včelařský, s.r.o.
    - License: Creative Commons Attribution-Share Alike 4.0 International license.
    - License Details: https://creativecommons.org/licenses/by-sa/4.0/deed.en
    - Accessed: May 10, 2024
    - Image Link: https://commons.wikimedia.org/wiki/File:European_foulbrood_CZ.jpg